Using Formative Assessment to Drive Mathematics Instruction in Grades PreK-2

Christine Oberdorf and Jennifer Taylor-Cox

Routledge
Taylor & Francis Group
LONDON AND NEW YORK

First published 2012 by Eye On Education

Published 2013 by Routledge
711 Third Avenue, New York, NY 10017, USA
2 Park Square, Milton Park, Abingdon, Oxon OX14 4RN

First issued in hardback 2015

Routledge is an imprint of the Taylor & Francis Group, an informa business

Cover Designer: Dave Strauss, 3FoldDesign

Library of Congress Cataloging-in-Publication Data

Oberdorf, Christine.
Using formative assessment to drive mathematics instruction in grades preK-2 / by Christine Oberdorf and Jennifer Taylor-Cox.
p. cm.
Includes bibliographical references.

1. Mathematics—Study and teaching (Early childhood) 2. Educational tests and measurements. 3. Mathematical ability—Testing. I. Taylor-Cox, Jennifer. II. Title.
QA11.2.O24 2011
372.7--dc23

2011012934

ISBN-13: 978-1-596-67187-4 (pbk)
ISBN-13: 978-1-138-13619-9 (hbk)

Also Available from EYE ON EDUCATION

Using Formative Assessment to Drive Mathematics Instruction in Grades 3–5
Christine Oberdorf & Jennifer Taylor-Cox

Family Math Night: Math Standards in Action
Jennifer Taylor-Cox

Family Math Night: Middle School Math Standards in Action
Jennifer Taylor-Cox & Christine Oberdorf

Math Intervention: Building Number Power with Formative Assessments, Differentiation, and Games (Grades PreK-2)
Jennifer Taylor-Cox

Math Intervention: Building Number Power with Formative Assessments, Differentiation, and Games (Grades 3-5)
Jennifer Taylor-Cox

Solving Behavior Problems in Math Class: Academic, Learning, Social, and Emotional Empowerment (Grades K-12)
Jennifer Taylor-Cox

Math in Plain English: Literacy Strategies for the Mathematics Classroom
Amy Benjamin

Engaging Mathematics Students Using Cooperative Learning
John D. Strebe

Mathematics Coaching Handbook: Working with Teachers to Improve Instruction
Pia M. Hansen

Differentiated Instruction for K-8 Math and Science: Ideas, Activities, and Lesson Plans
Mary Hamm & Dennis Adams

Buddies: Reading, Writing, and Math Lessons
Pia M. Hansen

A Collection of Performance Tasks and Rubrics: Primary School Mathematics
Charlotte Danielson & Pia M. Hansen

RTI Strategies that Work in the K-2 Classroom
Eli Johnson & Michelle Karns

About the Authors

Ms. Christine Oberdorf is a dedicated and experienced educator. She serves as a Staff Development Teacher for Montgomery County Public Schools in Maryland in addition to working as a consultant in mathematics education. She is a National Board Certified Teacher and currently serves on the Board of Directors of the Maryland Council of Teachers of Mathematics. Christine appreciates opportunities to enable educators as they empower young mathematicians. She lives in Arnold, Maryland with her husband Arthur and her children Robert and Aley.

Dr. Jennifer Taylor-Cox is an enthusiastic, captivating presenter and well-known educator representing Taylor-Cox Instruction, LLC. Jennifer serves as a consultant in mathematics education and classroom discipline, providing professional development opportunities for elementary, middle school, and high school educators in numerous districts across the United States. Her keynote speeches, professional development workshops, strategic coaching models, classroom demonstration lessons, study groups, and parent workshops are always high-energy and insightful.

Jennifer earned her Ph.D. from the University of Maryland and was awarded the "Outstanding Doctoral Research Award" from the University of Maryland and the "Excellence in Teacher Education Award" from Towson University. She served as the president of the Maryland Council of Teachers of Mathematics. Jennifer is the author of numerous professional articles and educational books.

Jennifer knows how to make learning mathematics engaging and meaningful, motivating learners of all ages! She has a passion for using formative assessments to differentiate math instruction and solve classroom discipline issues. Jennifer lives and has her office in Severna Park, Maryland. She is the mother of three children. Jennifer truly understands how to connect research and practice in education. Her zeal for improving the quality of education is alive in her work with educators, students, and parents.

If you are interested in learning more about the professional development opportunities Dr. Taylor-Cox offers, please contact her at www.Taylor-CoxInstruction.com or 410.729.5599.

Taylor-Cox
INSTRUCTION
Connecting Research and Practice in Education

Acknowledgements

Appreciation is expressed to…

Karren Schultz-Ferrell
Elementary Integrated Curriculum Instructional Specialist,
Montgomery County Public Schools
You are an expert in early childhood education and your feedback is invaluable.
Thank you for your friendship and encouragement.

Verna Washington
Instructional Specialist, Montgomery County Public Schools
You are always so willing to offer support. Thank you for sharing your knowledge and time.

Bonnie Ennis
Math Coordinator, Wicomico County Public Schools
Your perspective and feedback are greatly appreciated.

Joan Harvey
Retired Teacher, Montgomery County Public Schools
Your dedication to the education of children is a gift.
Thank you for your expertise and friendship.

Contents

CCSS denotes concepts emphasized in the Common Core State Standards at the PreK through second grade levels.

1

student resources

Introduction

achievement

how

what mathematics

What Is Formative Assessment?

Formative assessment drives mathematics instruction and is a key component in *Response to Intervention*. It is the process by which evidence of students' understanding is used by teachers to adjust instructional practice (Popham, 2008). As practitioners, we routinely monitor student performance on specific outcomes and standards. Formative assessments are employed to measure student performance so as to provide a targeted instructional response. Monitoring student learning through formative assessments provides a gauge, pinpointing where students are on the pathway of acquiring new knowledge. Their performance on these assessments provides work samples to analyze. The samples enable us to see where students are in comparison to where they need to be to meet the standard. Only through this process, are we equipped to provide an effective and meaningful instructional response. Without formative assessment, lesson planning is focused solely on curriculum with little regard for students' explicit academic needs.

How Does Formative Assessment Impact Student Achievement?

Effective formative assessment occurs simultaneously with instruction for the purpose of improving students' knowledge and performance in mathematics. When formative assessment is implemented properly, students learn what is being taught to a substantially greater degree (Black & Wiliam, 1998). When we provide feedback to students as a result of formative assessment, it is the most powerful factor in enhancing student achievement (Hattie & Jaeger, 1998). The National Mathematics Advisory Panel (2008) recommends regular use of formative assessment so that instruction can be adapted based on student progress. "Teachers' regular use of formative assessment improves their students' learning, especially if teachers have additional guidance on using the assessment to design and individualize instruction" (2008, p. xxiii). This book provides such guidance for teachers through a three-phase format of assessment, analysis, and response as illustrated below.

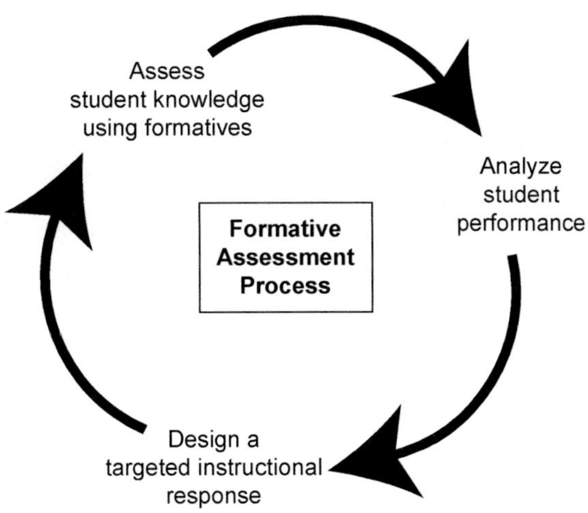

As practitioners, we experience tremendous pressure to "cover" the curriculum in a timely manner. Unfortunately, this sometimes translates to a practice of teaching *curriculum* rather than of teaching *children*. Teaching and learning form a dynamic alliance that is reliant on the interactions between teachers and students. These interactions serve as feedback to teachers and inform next steps for instruction designed to advance learning. To efficiently and effectively teach children, we must understand what they already know in order to plan meaningful next steps.

We have heard teachers lament about this process and become overwhelmed at the prospect of providing differentiated instruction for individual students. Advocating an individualized instructional program is neither realistic nor appropriate for most classroom teachers. All students are entitled to instruction designed to meet their identified needs, but this does not have to translate into a one-on-one instructional setting. Students can be grouped according to similar instructional needs. When analyzing student understanding of a math concept for an entire class, patterns and trends emerge and students' needs are often revealed in clusters. There may be times when we need to work with an individual student to reteach a concept or clear up a misconception; however, often there is a small group of students for whom the data show similar academic needs.

How Is This Book Organized?

Using Formative Assessment to Drive Mathematics Instruction contains seven chapters. This, the first chapter, identifies the purpose and intentions of the book by describing formative assessment and highlighting the impact of the process on student performance. Chapters 2 through 6 outline a process for the use of formative assessment to inform instruction. Each of these chapters addresses one of five content standards in mathematics: number and operations; algebraic thinking; geometry; measurement; and data analysis and probability. Within each content standard, key mathematics concepts are highlighted. These concepts were identified from such sources as the *Principles and Standards of School Mathematics* (NCTM, 2000), the *Curriculum Focal Points for Prekindergarten through Grade 8 Mathematics* (NCTM, 2006), and the *Common Core State Standards Initiative* (2010). The concepts directly addressing the Common Core State Standards are labeled as such in the table of contents and in the chapters with the symbol **CCSS**. Other concepts serve as foundations for later CCSS. Chapter 7 is a brief conclusion with final comments regarding the formative assessment process.

The formative assessment process in Chapters 2 through 6 is presented in a three-page format for each highlighted skill or concept. Each of the three pages is designed to answer the following questions regarding student performance and mathematics instruction:

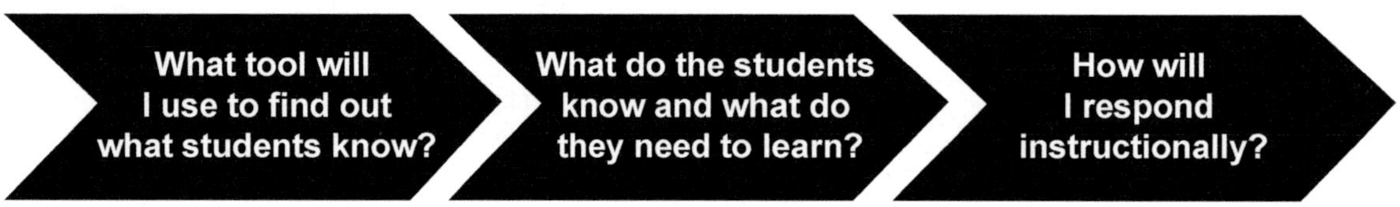

The formative assessments, student work samples, and suggested activities are provided for each mathematics concept to help teachers respond to these questions when planning instruction. Each is a deliberate step toward implementing effective mathematics instruction.

The **first** page in the three-page design illustrates a common sample of a *Traditional Formative Assessment* that one might find in a textbook or teacher resource (see Figure 1). Just below the assessment item is a *Limitations* note cautioning about potential shortcomings of the traditional assessment. The *Traditional Formative Assessment* is then followed by a suggested *Enhanced Formative Assessment,* which is provided as an alternative to the traditional format. The *Enhanced Formative Assessment* is designed to elicit responses offering more insight into student understanding and remedy the *limitations* associated with the traditional assessment.

Figure 1

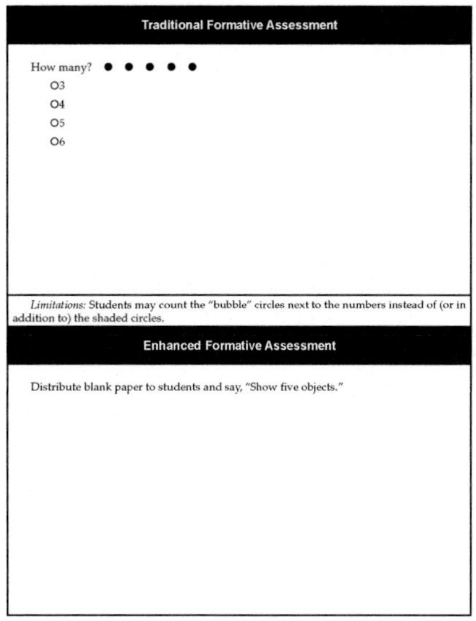

The **second** page in the three-page format includes four student responses to the enhanced formative assessment from the previous page (see Figure 2). The student responses are sequenced according to level of performance, beginning with a low level of understanding and ending with a high level of understanding for each concept. The first student work sample (top left in sample) demonstrates minimal understanding. The second and third student samples (top right and bottom left) demonstrate moderate understanding, but with varying degrees of success. The final work sample (bottom right) represents complete understanding of the assessed content. The shaded box just below each sample offers an analysis of the student work and guides the reader to the appropriate *Instructional Focus* provided on the next page.

Figure 2

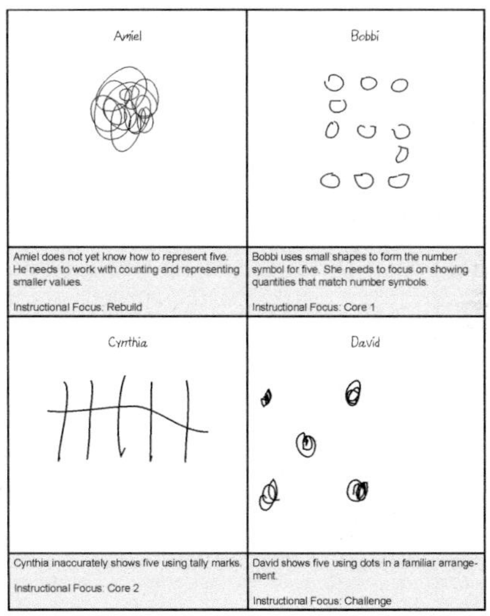

Although the instructional focus was developed with an individual student's response in mind, this is not to suggest that a teacher should develop or provide a different task for each student. It is not practical to expect teachers to implement individualized lessons. The student samples help to identify a level of understanding, in which several students may fall. The task is intended to address the needs of students with similar understanding who are working at a compatible level. The activities are designed to facilitate discourse among students in a small group setting.

The format of the **third** page (shown on next page) includes four small-group *Focus Activities* (see Figure 3). Each of the *Suggested Activities* is designed as a strategic response for students with instructional needs similar to those revealed in the work samples. The small-group *Rebuild Focus* is designed for students demonstrating minimal understanding. The activity introduces the key concepts through an interactive task and provides an opportunity to acquire new knowledge. The small-group *Core 1 Focus* and small-group *Core 2 Focus* activities are geared for students with moderate levels of understanding of the concept. Although both *Core Focus* tasks target students with some understanding of the content, each activity carries a different and specific focus based on an interpretation of needs from the student work samples. The *Core 1 Focus* activity is a slightly lower level of complexity than the *Core 2 Focus* activity. Lastly, the small-group *Challenge Focus* activity offers an enrichment opportunity for students demonstrating complete understanding of the math concept.

Each small-group *Focus Activity* is formatted with a defined goal, a list of the needed materials, a description and directions for the task, and potential questions to pose while students are engaged in the activity. Response to Intervention calls on teachers to provide effective and meaningful instruction. This resource provides a means to that end by serving as a guide for identifying students' needs and responding accordingly.

Figure 3

	Rebuild Focus	Core 1 Focus	Core 2 Focus	Challenge Focus
	Three or Four	Give Me Five	Tally Rally	Counting Counts
Goal	Count sets and represent values 3, 4	Show quantities that match number symbols 1–6	Use tally marks to represent values 1–6	Count sets and represent larger values (0–10)
Materials	♦ Counters (same size and same color) Trays or plates	♦ Number cube (die) Counters, Work mats, Dry erase board and marker	♦ Popsicle sticks Numeral cards (1–6)	♦ Numeral cards (0–10) Dot cards (0–10)
Suggested Activity Directions	Show 3 counters. Demonstrate counting by touching each object as it is counted aloud. Give each student 3 counters on a tray or plate. Ask students to take turns counting the objects. Encourage students to match the verbal number counted with the object touched. Repeat with 4 counters. Then give each student 3 or 4 counters to count. Students name the quantity and find others who have the same amount. Try greater values if students are ready.	Examine a number cube (die) and discuss the number of dots (1–6) on each side with students. Give each student 6 counters and a work mat. Roll the number cube and announce the number of dots shown. Students show the value using counters on individual work mats. Invite one student to write the number symbol on a dry erase board. Repeat. When a 5 is rolled, everyone stands up and gives one another a high 5 before showing the value with counters.	Give each student six popsicle sticks. Show the 1-4 number cards one at a time. Students show each value with popsicle sticks. The teacher helps the students arrange the popsicle sticks to resemble tally marks. Show the 5 card and explain how to place the "gate tally" diagonally across the four other sticks. Demonstrate 6 tally marks with popsicle sticks. After this introduction, shuffle the number cards and place facedown in a pile. Flip over the cards one at a time and have students show the values with popsicle sticks.	Shuffle number cards and dot cards. Place face down in an organized arrangement on the table. Students take turns flipping over 2 cards. If the quantity of dots matches the number symbol, the student keeps both cards. If there is not a match, cards are left face up and the next student flips over 2 more cards looking for a match. The activity continues until all cards are claimed. For further challenge, students can place unmatched cards facedown and try to remember where the cards are located.
Questions to Assess	♦ How many do you have? ♦ What number comes next?	♦ Is it five? How do you know? ♦ How could you show it another way?	♦ How many popsicle sticks do you need? ♦ Do you need a gate stick?	♦ What is the value? How do you know? ♦ Do the cards show the same amount?

How Can These Resources Be Utilized Effectively?

This book is designed to provide math educators with tools and resources targeting the varying needs of their students within five content standards of mathematics. When teaching a new concept or skill, students demonstrate a diverse range of competencies. A teacher introducing a lesson on addition, for example, must simultaneously consider those students who are uncertain what addition is, as well as those possessing a strong sense of addition and who are ready to tackle complex problems. The three-phase approach of this book assists teachers in identifying what students already know about the specific concept, facilitates meaningful student groupings according to current levels of understanding, and provides hands-on activities for instruction at four varying levels. It is not the intent of the authors that all students complete all four activities. Rather, the teacher may choose those activities most appropriate for groups of students.

Although the activities are designed to be an appropriate and targeted response to the identified needs of a student, some students may need to engage in more than one of the provided activities. The sequence of task difficulty suggests that students beginning at the *Rebuild Focus* level may well benefit from working through the *Core 1* and *Core 2* tasks as well, in order to meet the standard of performance measured in the *Enhanced Formative Assessment*. Additionally, the four activities do not dictate that a teacher must always manage four groups. Rather, the intent is to provide four instructional activities from which to choose to best meet the instructional needs of the class. The sequence offerred in the Table of Contents is not provided as a suggested order for teaching and should be adjusted to meet the specific scope and sequence of the district curriculum.

We offer you our best wishes as you navigate the journey to student success by *Using Formative Assessment to Drive Mathematics Instruction.*

2

Number & Operations

addition decompose numerals words numbers sets subtraction compose cardinal common place value fractions counting position

Traditional Formative Assessment

How many? ● ● ● ● ●

- O 3
- O 4
- O 5
- O 6

Limitations: Students may count the "bubble" circles next to the numbers instead of (or in addition to) the black circles.

Enhanced Formative Assessment

Distribute blank paper to students and say, "Show five objects."

CCSS Number & Operations: Counting Sets & Representing Values

Amiel does not yet know how to represent five. He needs to work with counting and representing smaller values.

Instructional Focus: **Rebuild**

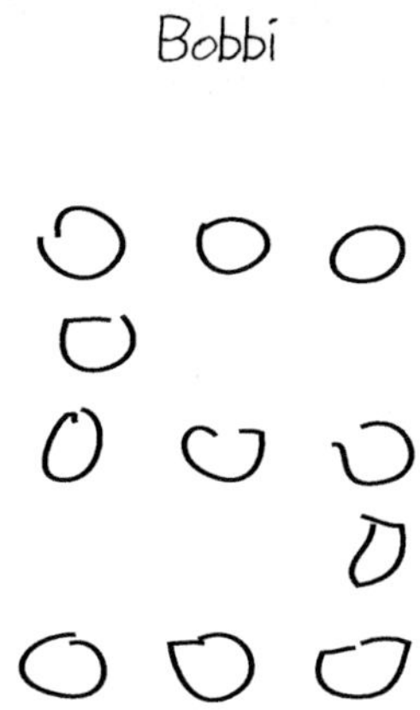

Bobbi uses small shapes to form the number symbol for five. She needs to focus on showing quantities that match number symbols.

Instructional Focus: **Core 1**

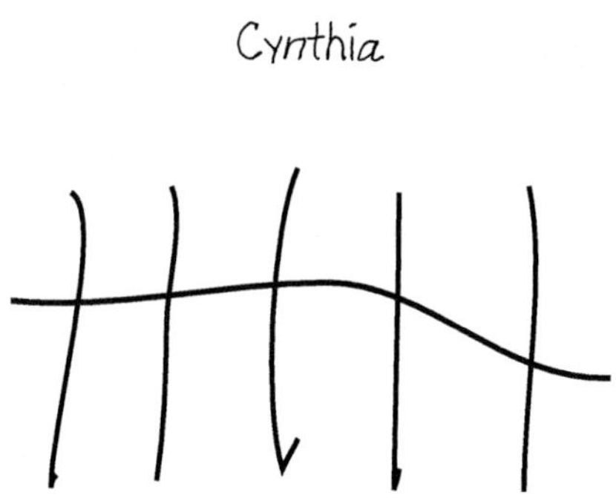

Cynthia inaccurately shows five using tally marks.

Instructional Focus: **Core 2**

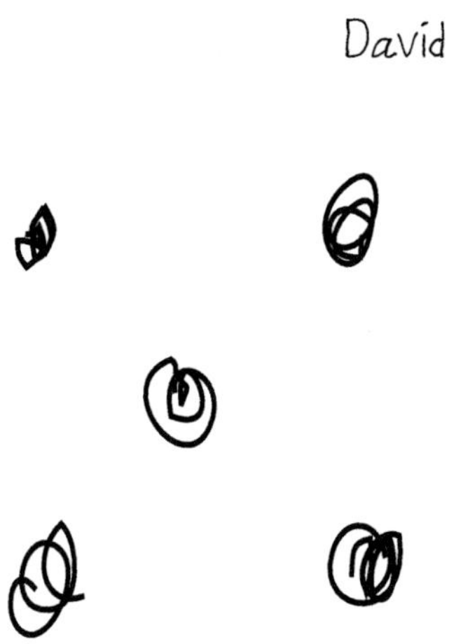

David shows five using dots in a familiar arrangement.

Instructional Focus: **Challenge**

CCSS Number & Operations: Counting Sets & Representing Values

	Rebuild Focus	Core 1 Focus	Core 2 Focus	Challenge Focus
	Three or Four	**Give Me Five**	**Tally Rally**	**Counting Counts**
Goal	Count sets and represent values 3, 4	Show quantities that match number symbols 1–6	Use tally marks to represent values 1–6	Count sets and represent larger values (0–10)
Materials	♦ Counters (same size and same color) ♦ Trays or plates	♦ Number cube (die) Counters ♦ Work mats ♦ Dry erase board and marker	♦ Popsicle sticks ♦ Numeral cards (1–6) 	♦ Numeral cards (0–10) ♦ Dot cards (0–10) 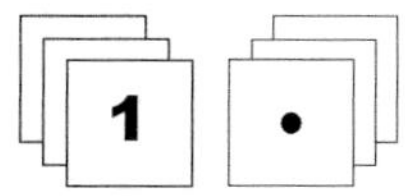
Suggested Activity Directions	Show 3 counters. Demonstrate counting by touching each object as it is counted aloud. Give each student 3 counters on a tray or plate. Ask students to take turns counting the objects. Encourage students to match the verbal number counted with the object touched. Repeat with 4 counters. Then give each student 3 or 4 counters to count. Students name the quantity and find others who have the same amount. Try greater values if students are ready. 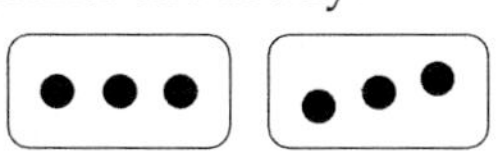	Examine a number cube (die) and discuss the number of dots (1–6) on each side with students. Give each student 6 counters and a work mat. Roll the number cube and announce the number of dots shown. Students show the value using counters on individual work mats. Invite one student to write the number symbol on a dry erase board. Repeat. When a 5 is rolled, everyone stands up and gives one another a high 5 before showing the value with counters. 	Give each student six popsicle sticks. Show the 1–4 number cards one at a time. Students show each value with popsicle sticks. The teacher helps the students arrange the popsicle sticks to resemble tally marks. Show the 5 card and explain how to place the "gate tally" diagonally across the four other sticks. Demonstrate 6 tally marks with popsicle sticks. After this introduction, shuffle the number cards and place facedown in a pile. Flip over the cards one at a time and have students show the values with popsicle sticks.	Shuffle number cards and dot cards. Place face down in an organized arrangement on the table. Students take turns flipping over 2 cards. If the quantity of dots matches the number symbol, the student keeps both cards. If there is not a match, cards are left face up and the next student flips over 2 more cards looking for a match. The activity continues until all cards are claimed. For further challenge, students can place unmatched cards facedown and try to remember where the cards are located.
Questions to Assess	♦ How many do you have? ♦ What number comes next?	♦ Is it five? How do you know? ♦ How could you show it another way?	♦ How many popsicle sticks do you need? ♦ Do you need a gate tally mark?	♦ What is the value? How do you know? ♦ Do the cards show the same amount?

Traditional Formative Assessment

Draw lines to match the numerals to the number words.

1	four
2	five
3	two
4	three
5	one

Limitations: Students may not be familiar with drawing lines to show answers.

Enhanced Formative Assessment

Distribute blank paper to students and say, "Show your favorite number in several ways. Try using the number word."

CCSS Number & Operations: Number Words & Numerals

Eva draws a picture of herself. She explains, "This is me when I was three."

Instructional Focus: **Rebuild**

Fletcher writes the numeral 3. Although he positions the symbol in several ways, he actually shows the number 3 in only one way.

Instructional Focus: **Core 1**

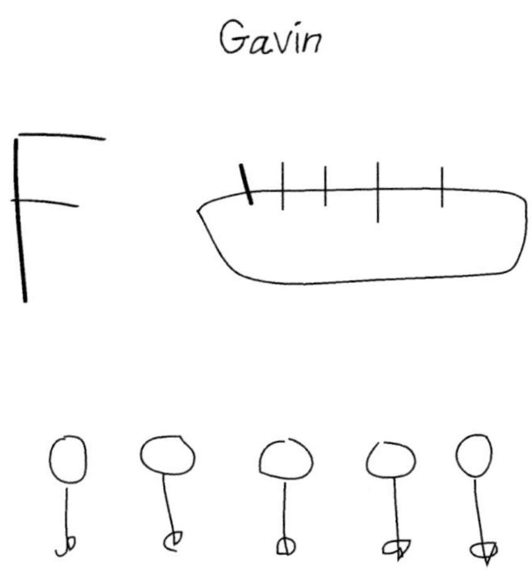

Gavin draws 5 flowers and 5 candles on a cake. He also writes the first letter in the word "five."

Instructional Focus: **Core 2**

Hyacinth writes the number symbol and the number word (phonetic spelling). She also draws a square and counts the corners using numerals.

Instructional Focus: **Challenge**

CCSS Number & Operations: Number Words & Numerals

	Rebuild Focus	Core 1 Focus	Core 2 Focus	Challenge Focus
	Shaving Cream Numbers	**Tracing Number Words**	**Number Word Fun**	**Number Mixup**
Goal	Write numerals (1–5)	Connect number words and numerals (1–3)	Connect number words and numerals (1–5)	Connect number words and numerals (1–10)
Materials	♦ Shaving cream ♦ Paper towels	♦ Number word templates (1–3) ♦ Dry erase markers ♦ Plastic page protectors	♦ Numeral cards (1–5) ♦ Number word cards (1–5) ♦ Dry erase boards and markers	♦ Number word cards (1–10) ♦ Dry erase boards and markers
Suggested Activity Directions	Demonstrate "air writing" by using the index finger to make 3 circles (counting aloud 1, 2, 3) and the number 3. Ask students to "air write" other numbers with you. After this introduction, spread a layer of shaving cream on the table. Help students write numerals and the corresponding number of circles. Encourage students to count aloud as they draw the circles and to say the number when writing the numeral. Try greater numbers if appropriate. 	Place number word templates in plastic page protectors. Students use dry erase markers to trace number words on the plastic sleeve. Students say the number as they write it. To build conceptual understanding, ask students to show the quantity with counters as they write and say the number words. To engage the students further, one student shows a quantity (1–3) of counters, while a partner traces and says the matching number. Try greater numbers if appropriate. **One**	Shuffle numeral cards and number word cards into two piles and place face-down. Flip over one card from each pile. Ask: Are they the same value? If the numeral and number word cards are not the same value, students write "No" on their dry erase boards. If the numeral and number word cards are the same, students write the numeral and number word on their dry erase boards. Continue through the pile of cards, shuffle cards, and repeat. **4** **Four**	Shuffle number word cards and place face-up in a line (forming a mixed-up number line). Ask students to think about how to put the number cards in the correct order (1–10). Students take turns describing which number card to move to a different location (to form an accurate number line). To indicate the number they want to move, students write the numeral and explain why that number word should be moved. Continue taking turns until all the number cards are in the correct order.
Questions to Assess	♦ How many circles? ♦ Does the numeral match the number of circles?	♦ How many counters? ♦ Which number word matches?	♦ What number is it? ♦ Does the numeral match the number word?	♦ Which number do you want to move? Why? ♦ Is the order correct yet?

Traditional Formative Assessment

Which sequence is correct?

- O 1, 2, 3, 5, 6, 7, 8, 9, 10
- O 1, 2, 3, 9, 5, 6, 7, 8, 4, 10
- O 1, 2, 3, 4, 5, 6, 7, 8, 9, 10
- O 1, 2, 3, 4, 5, 6, 7, 9, 10

Limitations: Students may not understand they need to look at each sequence individually.

Enhanced Formative Assessment

Count by ones. Fill in the missing numbers.

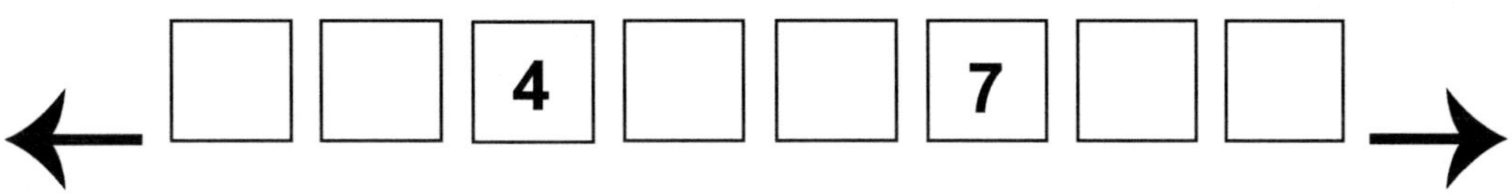

Ira

Count by ones. Fill in the missing numbers.

Ira draws smiley faces instead of writing the missing numbers.

Instructional Focus: **Rebuild**

Jolene

Count by ones. Fill in the missing numbers.

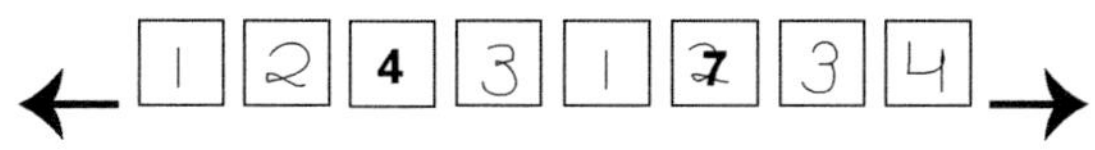

Jolene writes 1, 2, 3 in the available spaces. She begins again and even changes the "7" into a "2" to create 1, 2, 3, and 4.

Instructional Focus: **Core 1**

Kyle

Count by ones. Fill in the missing numbers.

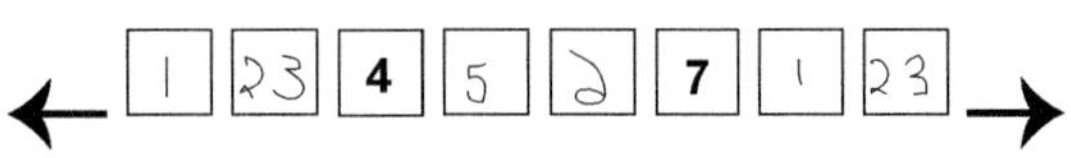

Kyle begins with 1 and squeezes 2 and 3 in the next box to maintain order. He does not accurately continue beyond 7.

Instructional Focus: **Core 2**

Lorraine

Count by ones. Fill in the missing numbers.

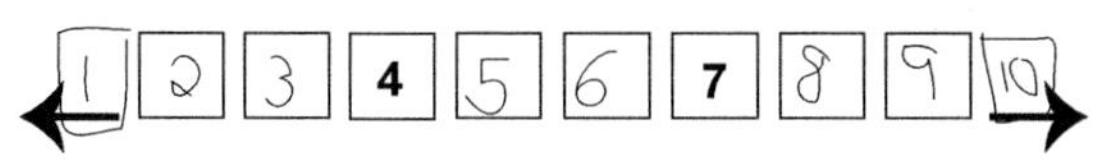

Lorraine accurately fills in the missing numbers. Additionally, she adds 1 and 10 in the appropriate spaces.

Instructional Focus: **Challenge**

CCSS Number & Operations: Relative Position of Whole & Cardinal Numbers

	Rebuild Focus	Core 1 Focus	Core 2 Focus	Challenge Focus
	Number Stomp	**Before and After**	**Magnetic Numerals**	**Number Line Chunks**
Goal	Build foundations in rote counting, number order, and cardinal numbers	Understand relative position of cardinal numbers (1–6)	Understand relative position of cardinal numbers (1–10)	Understand relative position of cardinal numbers (1–20) using number lines
Materials	♦ Timer	♦ Numeral cube (die labeled 1, 2, 3, 4, 5, 6) ♦ Before/After spinner	♦ Magnetic numerals	♦ Number lines (1–20) ♦ Mini-dry erase boards and markers
Suggested Activity Directions	Stand in a circle. Set the timer for approximately 5 seconds and allow a student to hold the timer. Together, everyone stomps their feet as they count aloud. When the timer rings, the student holding it says what number comes next. Pass the timer to another student, set it for a different number of seconds and repeat. Continue the activity allowing each child to have several turns.*	Demonstrate rolling the numeral cube and spinning the spinner. If the cube lands on 3 and the spinner lands on Before, announce, "Two is before three." Students take turns rolling the numeral cube, spinning the spinner, and announcing the number before or after the number rolled. 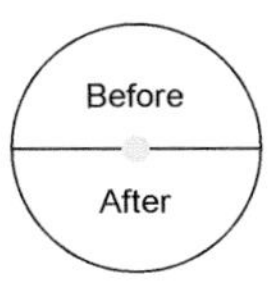 	Display magnetic numerals (1–6) in random order. Students take turns moving the numerals to place them in correct counting (by ones) order. Encourage students to explain how they know where the number should be positioned. Repeat including 7, 8, and 9. If appropriate, introduce 10. 	Give each student a number line (1–20). Prompt a discussion about the order of the numbers and that the arrows on the number line mean it keeps going forever. Demonstrate how to cover up portions of the number line with your hands and discuss the chunk of the number line that is visible. Ask students to draw number line chunks on dry erase boards. Encourage students to intentionally leave out numbers for partners to discover and name. 
Questions to Assess	♦ What number comes next? ♦ How did you know what number to say?	♦ What number did you roll? ♦ What number comes before/after?	♦ Where should you place the number? ♦ How do you know the number belongs there?	♦ Which chunk of the number line did you show? ♦ What are the missing numbers?

* A similar version of this activity appears in Taylor-Cox, *Math Intervention: Building Number Power with Formative Assessments, Differentiation, and Games* (Larchmont, NY: Eye On Education, 2009).

Traditional Formative Assessment
Circle the ways to make 10. 8+2 1+8 5+5 3+3+4 6+3 4+6 7+5
Limitations: Students may know how to compose and decompose numbers but they may not be familiar with how to communicate this with expressions.

Enhanced Formative Assessment
Show how to make 10 in several different ways.

Melinda

Show how to make 10 in several different ways.

Melinda draws the number symbol for 10 using two different fonts. She does not show how 10 is composed.

Instructional Focus: **Rebuild**

Ned

Show how to make 10 in several different ways.

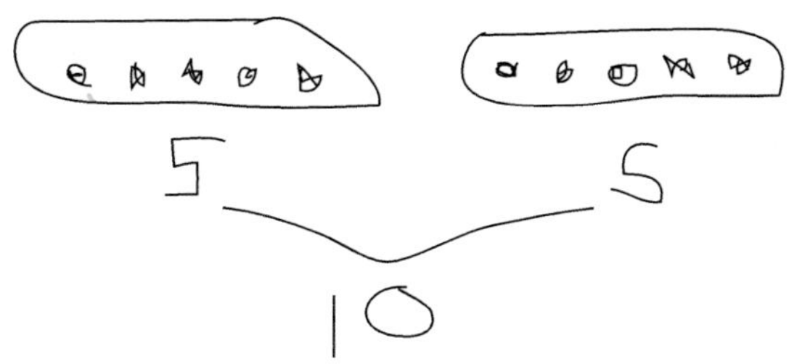

Ned shows how to compose 10 using two sets of 5. He does not show any other ways to compose 10.

Instructional Focus: **Core 1**

Osborn

Show how to make 10 in several different ways.

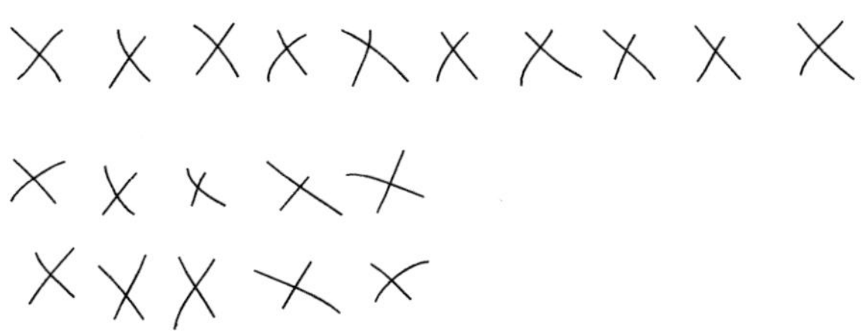

Osborn accurately shows two ways to compose 10.

Instructional Focus: **Core 2**

Paige

Show how to make 10 in several different ways.

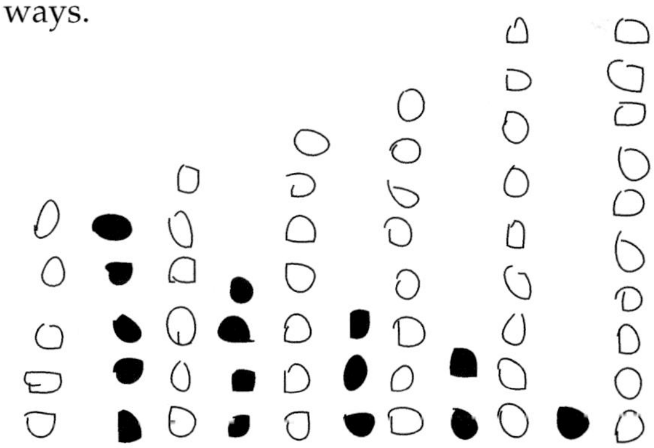

Paige accurately shows multiple ways to compose 10 using white and shaded shapes.

Instructional Focus: **Challenge**

CCSS Number & Operations: Compose & Decompose Numbers

	Rebuild Focus	Core 1 Focus	Core 2 Focus	Challenge Focus
	Towers of Five	**Towers of Ten**	**Ten-Frame Fun**	**Domino Display**
Goal	Compose and decompose 5	Compose and decompose 10 in several ways	Compose and decompose with more than two parts	Compose and decompose numbers (1–18)
Materials	♦ Linking cubes (same color)	♦ Linking cubes (sets of 10, same color)	♦ Ten-frames ♦ Colored chips (three colors)	♦ Set of double-nine dominoes
Suggested Activity Directions	Show students a tower of 5 linking cubes. Demonstrate how to break the tower into two parts. Discuss the parts and the total (e.g., 2 and 3 make a total of 5). Put the tower back together. Discuss and model other ways to break the tower of 5. After this introduction, break a tower of 5 and hide one of the parts under the table. Students look at the known part and try to figure out the hidden part. Repeat with different parts. Try other tower quantities if appropriate. 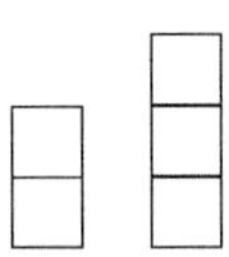	Model various ways to break a tower of 10 linking cubes into two parts. Partners try this activity by breaking a tower of 10 and hiding one part. The partner examines the part showing and names the hidden part. Ask students to record the parts and the total. Try other tower quantities if appropriate. 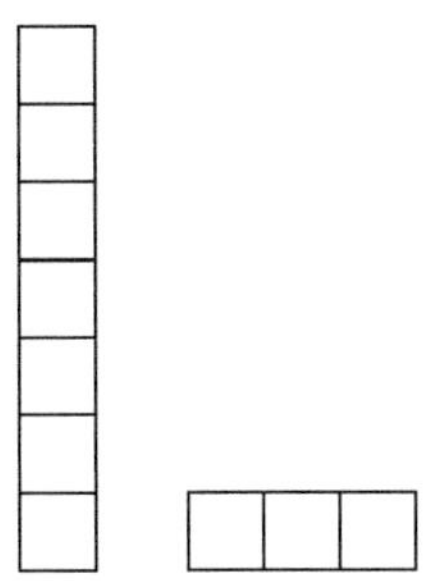	Show a 10-frame with 10 chips (same color). Discuss how the 10-frame is arranged. Show 10 chips (2 colors) on the 10-frame. Prompt discussion about how 10 is composed (example, 10 is made of 6 red chips and 4 blue chips). Encourage students to show 10 in different ways using chips of 2 colors. After practice, introduce composing 10 using 3 chip colors. 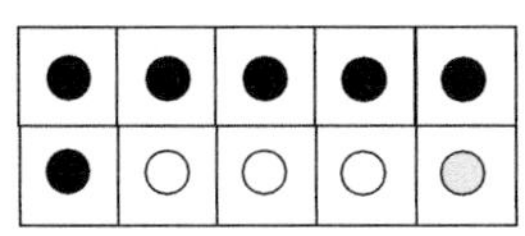	Display a set of double-nine dominoes. Choose a domino to examine. Discuss the dots in each region and the total (example, 5 dots and 7 dots make 12 dots). Ask students to look for other dominoes with 12 dots. Discuss each way 12 is composed (6 and 6, 5 and 7, 4 and 8, 3 and 9). Place these dominoes in a group. Find other domino quantities and make groups. Display the groups in order. 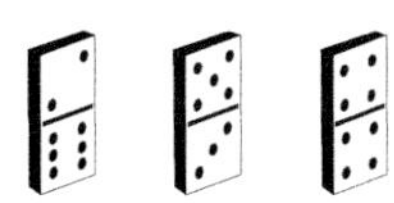
Questions to Assess	♦ What is the total? ♦ Can we break down five in another way?	♦ What is the hidden part? ♦ Do you know the total without counting?	♦ How did you show 10? ♦ How many parts of 10 could we have?	♦ What is the total number of dots? ♦ How many dominoes show 13 dots?

Traditional Formative Assessment

Circle the correct number.

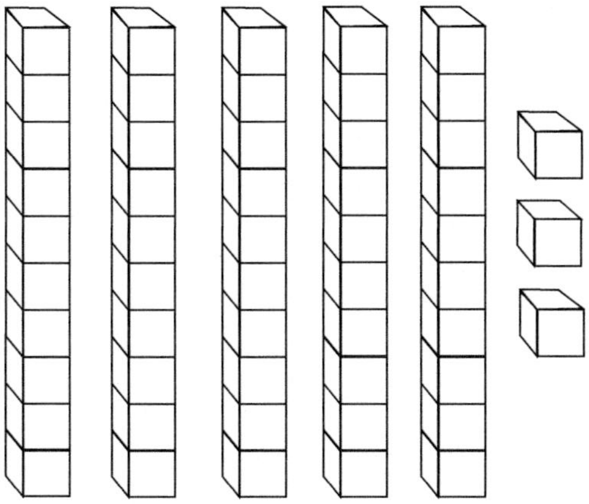

80   8   35   53   63

Limitations: Students may obtain correct answers (using counting by ones) without having an understanding of place value (counting by tens and ones).

Enhanced Formative Assessment

Distribute blank paper to students and say, "Write 53 and draw a model of 53."

CCSS Number & Operations: Basic Place Value

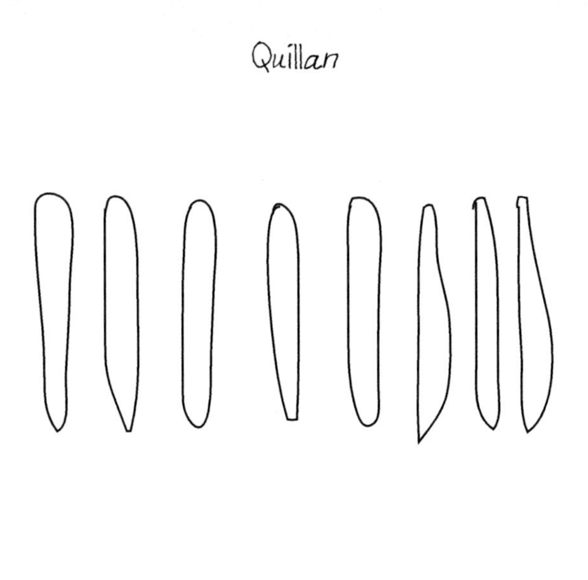

Quillan inaccurately draws 8 sticks (5 + 3) to represent 53.

Instructional Focus: **Rebuild**

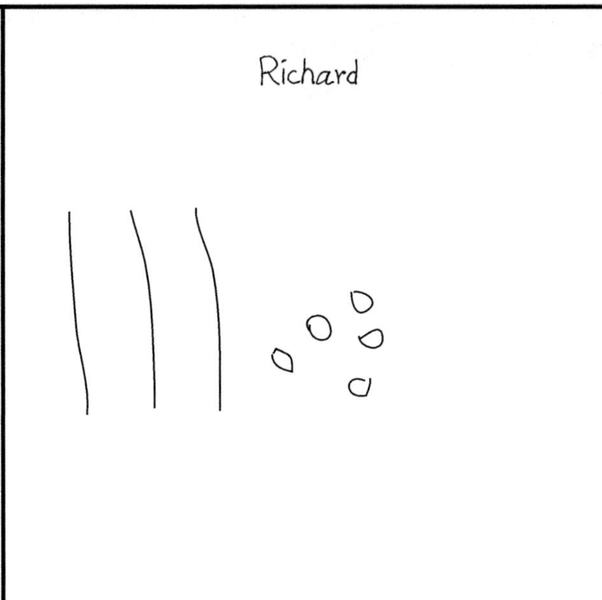

Richard shows tens and ones, but represents 35 instead of 53.

Instructional Focus: **Core 1**

Sandy accurately models 53 with 5 tens and 3 ones. She does not correctly write 53.

Instructional Focus: **Core 2**

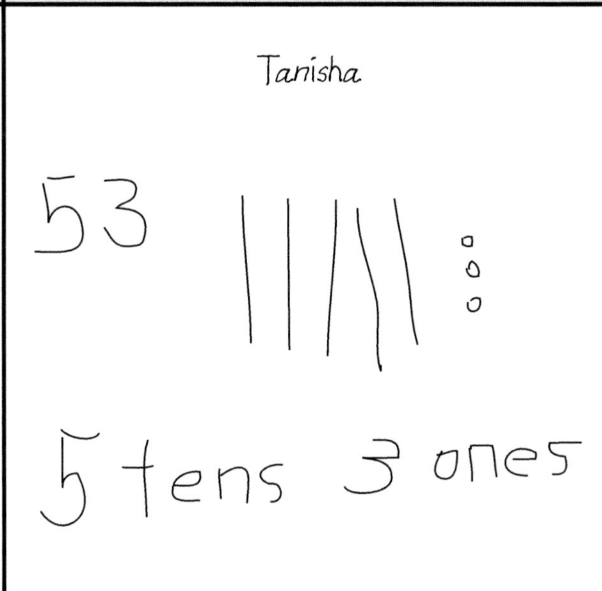

Tanisha correctly models and writes 53. She even explains how may tens and how many ones are in the number.

Instructional Focus: **Challenge**

CCSS Number & Operations: Basic Place Value

	Rebuild Focus	Core 1 Focus	Core 2 Focus	Challenge Focus
	Roll and Show	**Drop the Button**	**What's My Number?**	**Roll 3 Digits**
Goal	Model place value (1–20)	Model place value (1–99)	Use place value to write number symbols (0–99)	Model place value (3-digit numbers)
Materials	♦ Base 10 blocks ♦ Icosahedron die (20-sided)	♦ Base 10 blocks ♦ Place value mats ♦ Hundred chart (0–99) ♦ Button	♦ Base 10 blocks ♦ Place value mat ♦ Dry erase boards and markers	♦ Base 10 blocks ♦ Place value mats ♦ 3 Numeral cubes (labeled 1–6)
Suggested Activity Directions	Give each student 10 ones units. Ask students to show several numbers (one at a time in random order). Ask students to show 10 with units and introduce the 10 rod. Provide time for students to compare 10 units to one 10 rod (noting that the size is the same). Ask students to use the 10 rod and ones units to show values 11–19 (one at a time in random order). Introduce 20 (with two 10 rods, as 10 was introduced). Students take turns rolling the die while everyone displays the values with base 10 blocks.	Give each student a place value mat and some base 10 blocks. Drop the button on a 100 chart and announce the number where the button lands. Students show that number with base 10 blocks on place value mats. Help students show the number correctly and ask questions such as: How many tens? How many ones? Students clear their place value mats. Drop the button again to have a new number for students to model. Repeat.	Show a number with base 10 blocks and ask: What's my number? Students answer the question by writing the number on individual dry erase boards. Ask again: What's my number? Students show their written answers and say the answer. Corrections are made if needed. Continue the activity and allow students to take turns being the one to model a number and ask, "What's my number?" Include models with more than nine ones (e.g., 43 as 3 tens, 13 ones).	Introduce the 100 flat to students. Provide time for students to line up 10 base 10 rods on top of the 100 flat. Discuss how many tens/ones make the 100 flat. Roll 3 dice. Use one die for each digit in a 3-digit number. Students work with a partner to model the 3-digit number with base 10 blocks. Students take turns rolling the dice and modeling the 3-digit numbers.
Questions to Assess	♦ How many ones units do you need? ♦ Do you need a 10 rod? Why or why not?	♦ How many tens? ♦ How could you count the quantity by tens and ones?	♦ How many tens are in the number? ♦ Where did you write the ones digit?	♦ Where do the tens go? ♦ How could you write this number?

0	1	2	3	4	5	6	7	8	9
10	11	12	13	14	15	16	17	18	19
20	21	22	23	24	25	26	27	28	29
30	31	32	33	34	35	36	37	38	39
40	41	42	43	44	45	46	47	48	49
50	51	52	53	54	55	56	57	58	59
60	61	62	63	64	65	66	67	68	69
70	71	72	73	74	75	76	77	78	79
80	81	82	83	84	85	86	87	88	89
90	91	92	93	94	95	96	97	98	99

Tens	Ones

2 3 5

Hundreds	Tens	Ones

Traditional Formative Assessment

Circle the correct equations. Place X on the incorrect equations.

3 + 5 = 7   1 + 9 = 10   6 + 6 = 12   7 + 8 = 14

11 + 7 = 18   38 + 9 = 48   54 + 23 = 76   47 + 36 = 83

Limitations: Students may just guess as they circle or X the equations.

Enhanced Formative Assessment

What does it mean to add?

Solve:

7 + 8 =   38 + 9 =   54 + 23 =   47 + 36 =

Ulani

What does it mean to add? This is add +

Solve

7 + 8 = 8 38 + 9 = 9 54 + 23 = 23 47 + 36 = 36

Ulani does not offer a meaningful explanation of addition. All of her sums are incorrect.

Instructional Focus: **Rebuild**

Violet

What does it mean to add? to put toget er

Solve

7 + 8 = 15 38 + 9 = 48 54 + 23 = 50 47 + 36 = 76

Violet provides an accurate explanation of addition (to put together). She accurately solves the single-digit addition problem. She adds 10 instead of 9 on the second problem and does not accurately solve the third and fourth problems.

Instructional Focus: **Core 1**

Whitman

What does it mean to add? cowt up

77 78
67
57

Solve

7 + 8 = 15 38 + 9 = 47 54 + 23 = 76 47 + 36 = 78

48 − 1 64
74 75 76

Whitman's explanation of addition (count up) is solid. He correctly solves the first and second problems. He attempts to use the partial sums strategy to solve the other problems.

Instructional Focus: **Core 2**

Xeno

What does it mean to add? join

Solve

7 + 8 = 15 38 + 9 = 47 54 + 23 = 77 47 + 36 = 83

70 + 7 70 + 13
77 83

Xeno's explanation of addition is accurate. He successfully solves all of the problems. His use of the partial sums strategy is apparent.

Instructional Focus: **Challenge**

CCSS Number & Operations: Addition

	Rebuild Focus	Core 1 Focus	Core 2 Focus	Challenge Focus
	Number Stories	**Nine Is Close to Ten**	**Naming Partial Sums**	**Some Sums**
Goal	Single-digit addition in context (number stories)	Adding 9 using the near 10 strategy	Adding two-digit numbers using partial sums strategy	Adding three-digit numbers
Materials	♦ Counters ♦ Plates ♦ 2 decahedron dice (10-sided, labeled 0–9)	♦ Hundred charts (1–100) ♦ Transparent chips ♦ Dry erase board and marker	♦ 4 sets digit cards (0–9) ♦ Addition template ♦ Dry erase boards and markers	♦ Decahedron die (10-sided) ♦ Addition template (3-digit)
Suggested Activity Directions	Roll the decahedron dice. Show the amounts with sets of counters on a plate. Tell a number story that requires adding the two numbers and ask a question about the sum (e.g., I had 7 cookies. Maria gave me 3 more cookies. How many cookies do I have now?) Students name the sum and the equation (7 + 3 = 10). Students take turns rolling, modeling, and telling number stories for the group to solve. 	Drop the chip onto the hundred chart. Write an equation using the number as one addend and 9 as the other addend on the dry erase board (example, 38 + 9 = ?). Students place a chip on 38 on their hundred chart and move the chip down (plus 10) and over (minus 1). Write 38 + 9 = 48 – 1. Students take turns dropping the chip, writing, and modeling adding 9 (by using 10) to a 2-digit number. 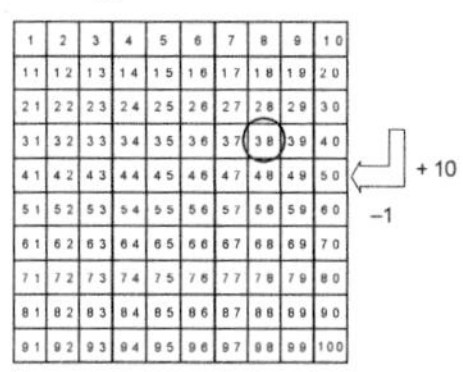	Shuffle each set of digit cards and place face-down in a pile. Flip over one card from each pile and place these cards in the addition template. Students write the equation on their dry erase boards. Students write and say the partial sums (example, 47 + 36 = 70 + 13). Help students focus on the partial sums that make the total sum.* 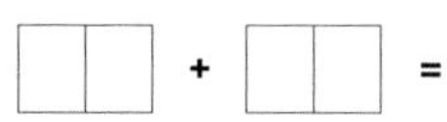	Give each student an addition template. Roll the decahedron die. Students decide where to write the digit on the template. Continue rolling the decahedron die for a total of 6 times. Students decide where to write digits after each roll. Once a digit is placed, it cannot be changed. Students add and compare the totals to see which total is greater. 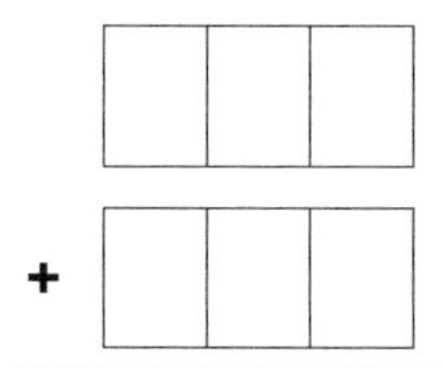
Questions to Assess	♦ How many did you start with? ♦ How many more did you receive?	♦ What is the equation? ♦ How can you use plus 10 to help you solve plus 9?	♦ What are the partial sums? ♦ Does it matter if you add the tens first?	♦ Where do you want to place the digit? Why? ♦ How do you add hundreds?

* A similar version of this activity appears in Taylor-Cox, *Math Intervention: Building Number Power with Formative Assessments, Differentiation, and Games* (Larchmont, NY: Eye On Education, 2009).

Traditional Formative Assessment

You have 8 minutes to solve these problems.

7 − 5 =   12 − 6 =   50 − 10 =   70 − 30 =

68 − 9 =   72 − 40 =   98 − 36 =   83 − 54 =

Limitations: Some students may know how to subtract, but fail to respond correctly because they are intimidated by timed tests.

Enhanced Formative Assessment

Some of the problems are incorrect. Fix the errors.

10 − 4 = 6   60 − 10 = 70   83 − 60 = 23

$$\begin{array}{r} 27 \\ -\ 9 \\ \hline 17 \end{array} \qquad \begin{array}{r} 76 \\ -\ 25 \\ \hline 51 \end{array} \qquad \begin{array}{r} 92 \\ -\ 73 \\ \hline 21 \end{array}$$

Yolanda

Some of the problems are incorrect. Fix the errors.

10 + 4 = ~~6~~ 14 60 + 10 = 70 83 + 60 = ~~23~~

27	76	92
+ 9	+ 25	+ 73
~~17~~	~~51~~	~~21~~

Yolanda changed the subtraction sign to addition sign in all of the problems.

Instructional Focus: **Rebuild**

Zach

Some of the problems are incorrect. Fix the errors.

Good 10 – 4 = 6 Good 60 – 10 = 70 Bad 83 – 60 = 23

Bad	Bad	Good
27	76	92
– 9	– 25	– 73
~~17~~ 18	51	21
𝍸 IIII		

Zach knows the first problem is correct. He misses the second and third problems. He accurately corrects the fourth problem (using counting down by ones) and misses the fifth and sixth problems.

Instructional Focus: **Core 1**

Amber

Some of the problems are incorrect. Fix the errors.

10 – 4 = 6 √ 60 – 10 = 70 50 83 – 60 = 23 √

27	76	92
– 9	– 25	– 73
17 + 1	51	21
18	√ √	√ ?

Amber correctly solves all of the problems except the last one.

Instructional Focus: **Core 2**

Bruce

Some of the problems are incorrect. Fix the errors.

10 – 4 = 6 60 – 10 = ~~70~~ 50 83 – 60 = 23

27	76	80 12 ~~92~~
– 9	– 25	– 73
~~17~~ 18	51	~~21~~ 19

Bruce accurately solves all of the problems. He is ready to learn other strategies that will strengthen his mental math abilities.

Instructional Focus: **Challenge**

CCSS Number & Operations: Subtraction

	Rebuild Focus	Core 1 Focus	Core 2 Focus	Challenge Focus
	Part-Part-Whole	**Minus 10**	**Regroup or Not?**	**Subtraction Action**
Goal	Build understanding of subtraction with part-part-whole strategies	Subtract 10 and multiples of 10	Subtract 2-digit numbers with and without regrouping	Subtract using the partial differences strategy
Materials	♦ Chips (red/yellow)	♦ Base 10 blocks ♦ Hundred chart (1–100)	♦ Base ten blocks ♦ Dry erase boards and markers	♦ Dry erase boards and markers ♦ Number lines (–9 through +9)
Suggested Activity Directions	Model part-part-whole relationships with chips (yellow part 4 + red part 5 = total 9). Say: "If you know the number of yellow chips and the total number of chips, how could we figure out how many red chips there are? Encourage students to share how to find the unknown part. Give a different total and known part. Ask students to use chips to model solving the unknown part. Write the corresponding subtraction equation. Discuss how the minus sign looks differently from the addition sign. Try other totals and parts. ○○○○ ●●●●●	Encourage students to skip count by tens forward and backward. Discuss how backward skip counting represents subtraction. Ask students to show 60 with base 10 blocks. Point to 60 on a hundred chart. Ask students to count by tens backward as they take away 10 rods from the model. Point to multiples of 10 on the hundred chart as students subtract. Repeat beginning with 90. When students are ready, use the same strategies to subtract multiples of 10 from any 2-digit number (example, 42 – 30 = count, 42, 32, 22, 12).	Write 74 – 23. Ask students to model solving this equation with base 10 blocks. Highlight the fact that there is no need to regroup. Then try 74 – 26. Model and discuss trying to take 6 ones from 4 ones. Ask students to show 74 in two different ways (7 tens and 4 ones, 6 tens and 14 ones) with base 10 blocks. Discuss how both models show 74. Encourage students to notice the simplicity of subtracting 6 ones from 14 ones. Highlight this as subtraction with regrouping. Give students several subtraction problems to decide if regrouping is needed or not.	Distribute number lines (–9 through +9). Discuss how negative numbers have "owe" values. Ask students to write 92 – 73 =. Ask students to find the partial difference in the tens (20) and use the number line to find the partial difference in the ones (–1). Ask students to combine the differences (20 – 1 = 19). Try other subtraction situations with and without regrouping. If students are ready, introduce 3-digit subtraction. *NOTE:* The number line can be used for the tens, if counting by tens.
Questions to Assess	♦ What is the total? ♦ What is subtraction?	♦ What is 10 less? ♦ How did you find the difference?	♦ Is regrouping needed? Why or why not? ♦ How many ones/tens?	♦ What does negative mean? ♦ How does the number line help?

Traditional Formative Assessment

How much is shaded?

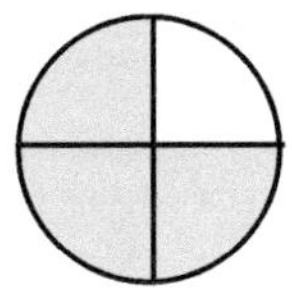

$\frac{1}{4}$ $\frac{3}{4}$ $\frac{1}{3}$ $\frac{4}{3}$

Limitations: Some students may know fourths, but not understand that shaded means the darker parts.

Enhanced Formative Assessment

Draw a picture to show each fraction.

$\frac{1}{4}$ $\frac{3}{4}$ $\frac{1}{3}$ $\frac{4}{3}$

CCSS Number & Operations: Basic Common Fractions

Charlie

Draw a picture to show each fraction.

$\frac{1}{4}$ $\frac{3}{4}$ $\frac{1}{3}$ $\frac{4}{3}$

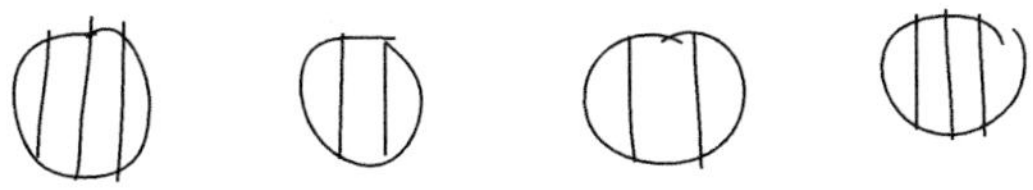

Charlie's fractions are not accurate because the parts are not equal size and he did not otherwise indicate the fractional parts.

Instructional Focus: **Rebuild**

Donna

Draw a picture to show each fraction.

$\frac{1}{4}$ $\frac{3}{4}$ $\frac{1}{3}$ $\frac{4}{3}$

Donna accurately shows the fourths, but does not correctly show the thirds.

Instructional Focus: **Core 1**

Evan

Draw a picture to show each fraction.

$\frac{1}{4}$ $\frac{3}{4}$ $\frac{1}{3}$ $\frac{4}{3}$

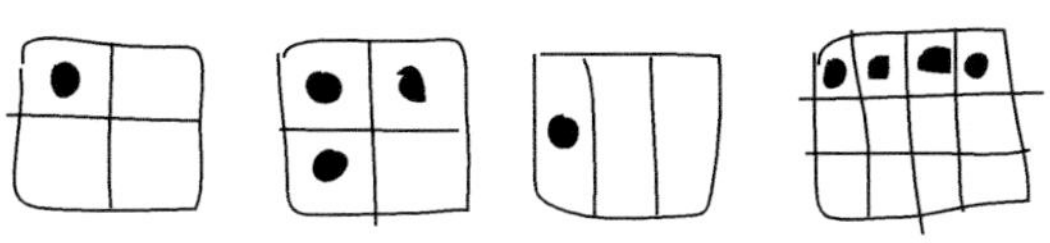

Evan's pictures correctly show the first 3 fractions. He does not know how to show improper fractions.

Instructional Focus: **Core 2**

Faith

Draw a picture to show each fraction.

$\frac{1}{4}$ $\frac{3}{4}$ $\frac{1}{3}$ $\frac{4}{3}$

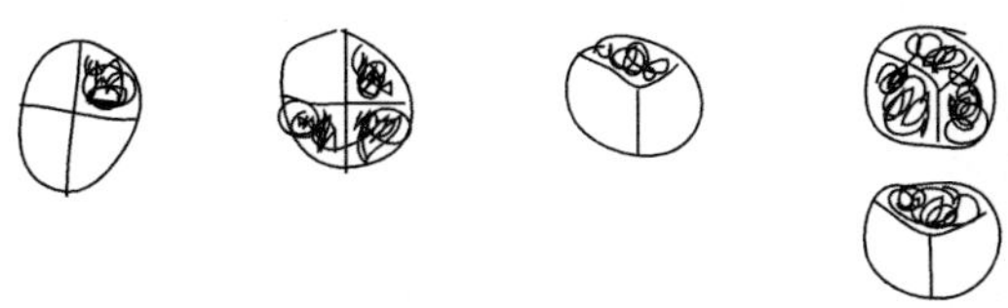

Faith accurately draws all of the fractions.

Instructional Focus: **Challenge**

CCSS Number & Operations: Basic Common Fractions

	Rebuild Focus	Core 1 Focus	Core 2 Focus	Challenge Focus
	Have Any Halves?	**Rectangle Fractions**	**Show the Fraction**	**Clothesline Fractions**
Goal	Represent halves (equal parts)	Represents thirds and fourths (and other fractions)	Represent fractions greater than one (improper fractions)	Order proper and improper fractions
Materials	♦ Variety of paper shapes (different colors) cut into two parts (halves and not halves) ♦ Dry erase boards and markers	♦ Geoboards and Geobands	♦ Fraction Bars ♦ Fraction Circles ♦ 2 Decahedron die (different colors)	♦ Clothesline ♦ Clothespins ♦ Index cards ♦ Markers ♦ Numeral dice (1–6, two different colors)
Suggested Activity Directions	Mix up shape parts. Students look through parts to make whole shapes. After all shape parts are matched, ask students to sort by halves and not halves. Students explain why a shape is or is not divided in half (equal parts versus unequal parts). Students draw representations of halves and not halves on dry erase boards. Try fourths; then thirds when students are ready. 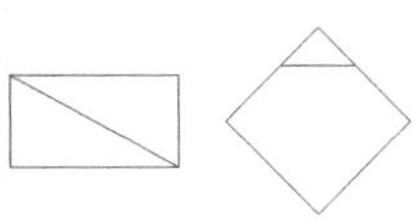	Students show a rectangle divided into fourths with geobands on geoboards. Discuss rectangle sizes, how the fourths are equal, and how many geobands are used. Ask students to show a rectangle with three equal parts (thirds). Check that everyone has three equal parts. Discuss the number of geobands used for thirds compared to the number used for fourths. Repeat with sixths and eighths of rectangles. 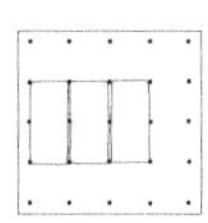	Designate one decahedron die for the numerator and the other die for the denominator. Roll the dice and write the fraction. Students represent the fraction using fraction bars or fraction circles. Students compare and verify the representations. Encourage students to decide if each fraction formed is an improper fraction (numerator is equal to or greater than the denominator) or a proper fraction (numerator is less than the denominator).	Hang string to resemble a clothesline. Students take turns rolling the dice. One number is the numerator, the other is the denominator. Students write each fraction on an index card and hang it, in order, on the clothesline (serving as a number line). Adjust the placement of each fraction card if needed. Encourage students to draw the fractions on the cards, as well.* 
Questions to Assess	♦ What does half mean? ♦ What if there were four equal parts?	♦ Are the parts equal? ♦ Did you use an odd or even number of geobands?	♦ How will you represent the fraction? ♦ Is it a proper fraction? Why or why not?	♦ Which number is the denominator? ♦ Is it an improper fraction?

* A similar version of this activity appears in Taylor-Cox, *Family Math Night: Math Standards in Action* (Larchmont, NY: Eye On Education, 2005).

3

symbols reasoning

solving

Algebraic Thinking

growing cummutativity problem

repeating

equality balance patterns

Traditional Formative Assessment

What is next in the pattern?  ______

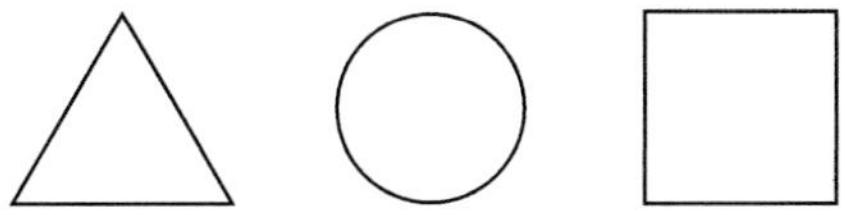

1 2 3

Limitations: The unit in this repeating pattern is actually unknown because it is not repeated.

Enhanced Formative Assessment

Note: This assessment can be modeled with objects or completed on paper.

Continue the repeating pattern.

 ______ ______ ______

Create a different repeating pattern.

______ ______ ______ ______ ______ ______ ______ ______

Algebraic Thinking: Repeating Patterns

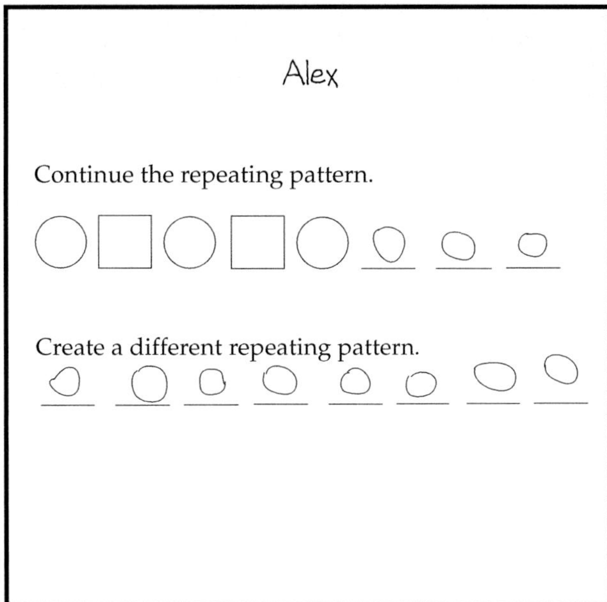

Alex uses his one-to-one correspondence skills to accurately draw a shape on each line. He uses one of the shapes (circle) from the pattern unit. Alex does not continue or create a different pattern correctly because he uses only circles.

Instructional Focus: **Rebuild**

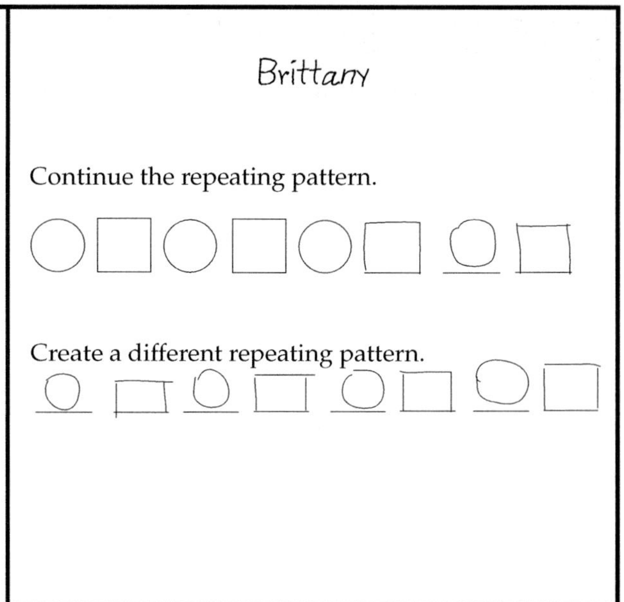

Cleverly, Brittany incorporates the line into her drawings of squares in the pattern. She accurately continues the pattern. Brittany does not create a different pattern; she uses the original pattern.

Instructional Focus: **Core 1**

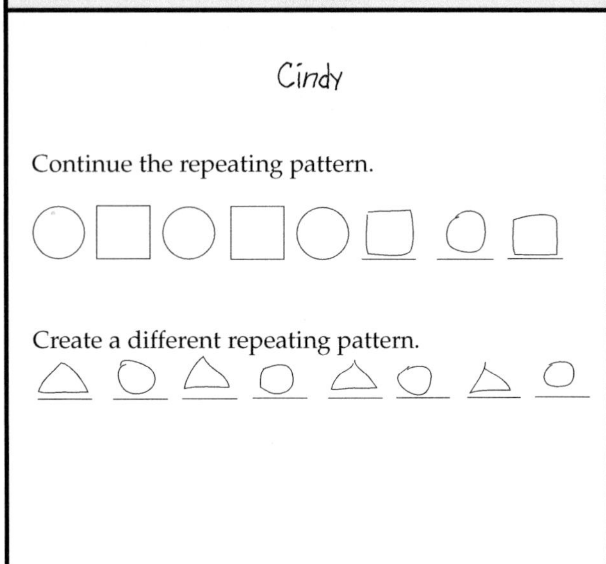

Cindy continues and accurately creates a different AB pattern using the correct shapes. She is ready to learn about more complex patterns.

Instructional Focus: **Core 2**

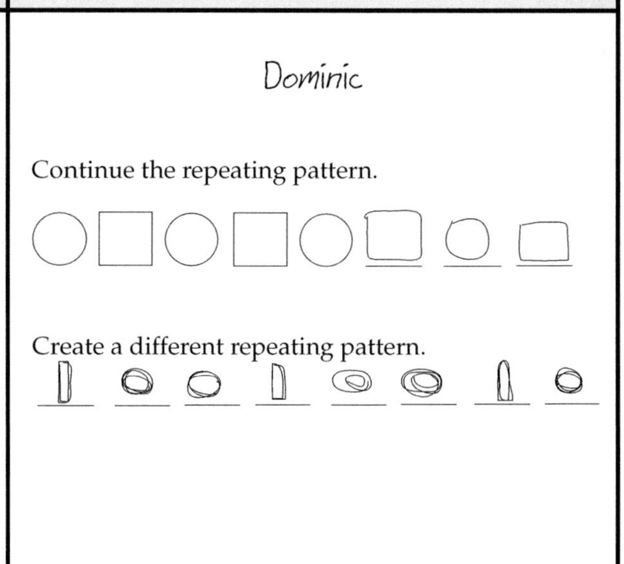

Dominic accurately continues the pattern. He moves beyond AB patterns by creating a complex ABB pattern.

Instructional Focus: **Challenge**

Algebraic Thinking: Repeating Patterns

	Rebuild Focus	Core 1 Focus	Core 2 Focus	Challenge Focus
	Color Patterns	**Shape Patterns**	**Patterns with Bears**	**Xs and Os**
Goal	Continue and describe basic repeating patterns	Create and translate basic repeating patterns	Continue and describe more complex repeating patterns	Create and translate more complex repeating patterns
Materials	♦ Color cubes ♦ Sentence strip	♦ Pattern blocks ♦ Crayons ♦ Paper	♦ Colored counting bears	♦ Pattern blocks ♦ Dry erase boards and markers
Suggested Activity Directions	Show students a pile of cubes (2 colors). Ask students to sort the cubes by color. Demonstrate how to make a pattern by taking one cube at a time from alternating piles and lining the cubes up. Say each color as the pattern is made. After the repeating unit has been revealed several times, ask students which color comes next and why. Encourage students to create their own color patterns with the cubes in the same manner. If appropriate, students draw the patterns they created on sentence strip. 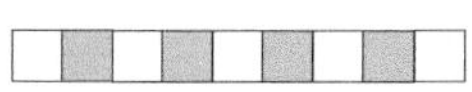	Use pattern blocks to create a repeating AB pattern for students to examine. Demonstrate how to create the same type of pattern with different pattern blocks by matching one pattern block to another (e.g., hexagons matched to trapezoids and squares matched to triangles). Encourage students to try translating the AB pattern with different pattern blocks. When students are ready, introduce translating an AB pattern into a color pattern created with crayons. If appropriate, ask students to illustrate patterns.	Make a repeating ABB pattern with the colored counting bears. Ask students to examine the pattern and describe which color bear comes next. Provide time for students to use counting bears to make ABB patterns. Students ask one another which bear comes next in the pattern. Introduce using the same color bear in different positions (e.g., up, down, down, up, down, down). If students are ready, continue other types of repeating patterns (e.g., AAB, ABC, AABB, ABBB).	Show students a repeating pattern such as AAB with pattern blocks. Demonstrate how to translate the pattern using Xs and Os written on a dry erase board. Discuss how both patterns are alike. Try other types of patterns with the pattern blocks. Ask students to translate the patterns into patterns with Xs and Os. Invite students to think of other ways to translate the patterns (using different symbols). XXOXXOXXO
Questions to Assess	♦ Which color cube comes next? ♦ What is the repeating pattern core? ♦ How can you describe the repeating pattern?	♦ How many things are in the core of this repeating pattern? ♦ How can you translate this pattern?	♦ How do you know this is a pattern? ♦ How many different things are in the pattern core?	♦ How can you compare the patterns? ♦ What makes these patterns the same?

Traditional Formative Assessment

Continue the pattern.

1, 3, 5, ___, ___, ___

Limitations: Because "growing" pattern is not specified, some students may create a repeating pattern. Additionally, showing the values with number symbols rather than with pictures or objects at this early stage may actually mask the algebraic understanding of a student who may not yet read and understand number symbols.

Enhanced Formative Assessment

Note: This assessment can be modeled with objects or completed on paper. Record students' descriptions.

Continue the growing pattern.

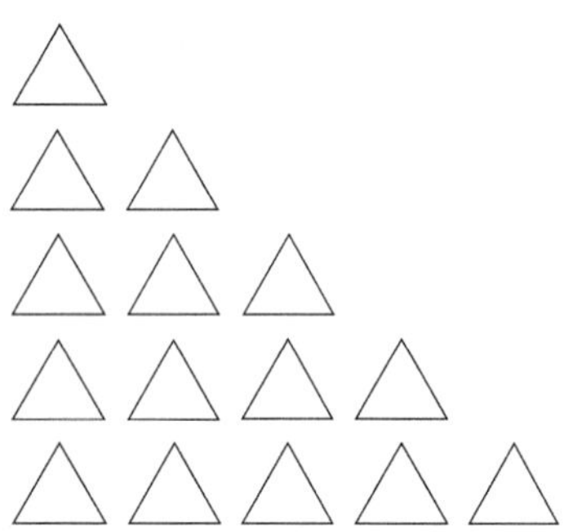

Describe the growing pattern.

CCSS Algebraic Thinking: Growing Patterns

Erica

Continue the growing pattern.

Describe the growing pattern. *It is all filled up now.*

Erica adds triangles to the pattern to create an array. She does not continue the growing pattern. Her description is related to her array but unconnected to the linear growing pattern.

Instructional Focus: **Rebuild**

Freddie

Continue the growing pattern.

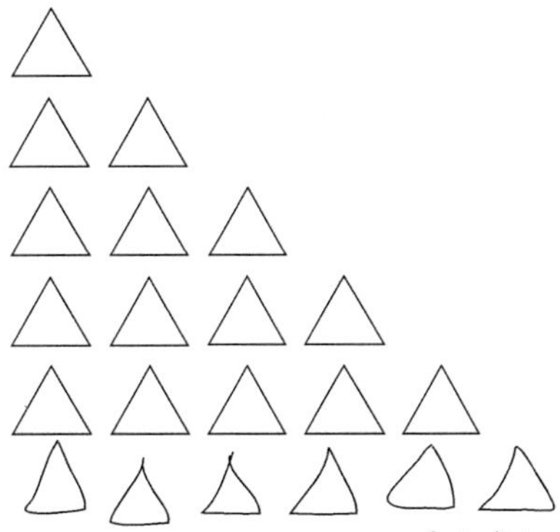

Describe the growing pattern. *I don't know.*

Freddie uses the correct number of triangles in the next term in the growing pattern. He can continue the pattern. He does not provide an accurate description of the growing pattern.

Instructional Focus: **Core 1**

Gabriel

Continue the growing pattern.

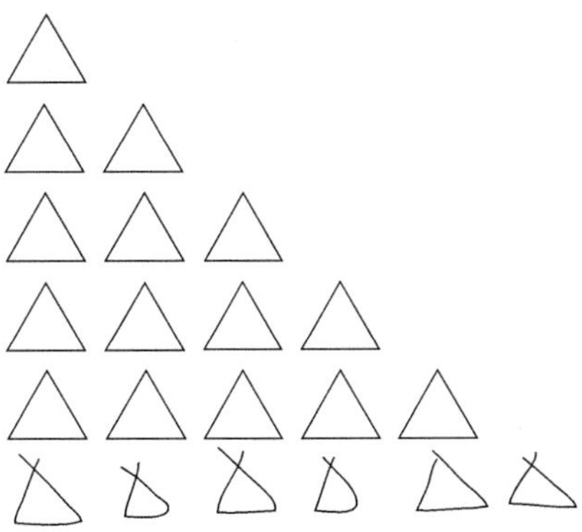

Describe the growing pattern. *Another row.*

Gabriel continues the growing pattern using the correct number of triangles. His explanation of the growing pattern shows he has partial understanding of how to describe the growing pattern.

Instructional Focus: **Core 2**

Hannarose

Continue the growing pattern.

Describe the growing pattern. *There's one more every time. It can keep going.*

Hannarose accurately adds two more terms to the growing pattern. Her description reveals her high level of understanding.

Instructional Focus: **Challenge**

	Rebuild Focus	Core 1 Focus	Core 2 Focus	Challenge Focus
	Grow by 1 Patterns	**Upstairs Downstairs**	**Chip Patterns**	**Tricky Patterns**
Goal	Continue basic (+1) growing patterns	Describe basic (+1) growing patterns	Continue and describe more complex growing patterns (starting with zero)	Create more complex growing patterns (starting with values other than zero)
Materials	♦ Color tiles (1 inch) ♦ Graph paper (inches) ♦ Crayons	♦ Connecting cubes	♦ Counting chips	♦ Connecting cubes ♦ Dry erase boards and markers
Suggested Activity Directions	Show a growing pattern (1, 2, 3, 4, 5, . . .) with color tiles on the graph paper using tiles of the same color. Ask students to name and show the next term in the pattern. Ask students to show the same growing pattern with tiles of a different color. Continue the patterns by adding the next few terms. Demonstrate how to show the same pattern by coloring the graph paper using matching colors. If students are ready, use mixed colors to show other "grow-by-one" patterns.	Connect cubes to make 5 towers that consecutively increase by 1 block. Place towers in order (1, 2, 3, 4, 5). Invite students to see how the tops of the towers look like stairs. Describe moving up and down the stairs. Encourage students to notice the equal change (±1). Ask students to create and describe their own growing patterns using the connecting cubes. 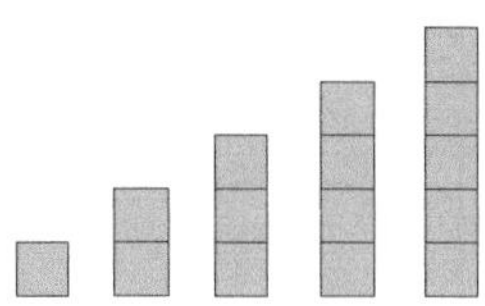	Demonstrate a simple "grow-by-one" pattern with counting chips. Explain how growing patterns can "grow" by different consistent amounts. Encourage students to use the counting chips to create and describe "grow-by-two" patterns. Try "grow-by-three" and "grow-by-four" patterns. Provide time for students to continue and describe one another's growing patterns. 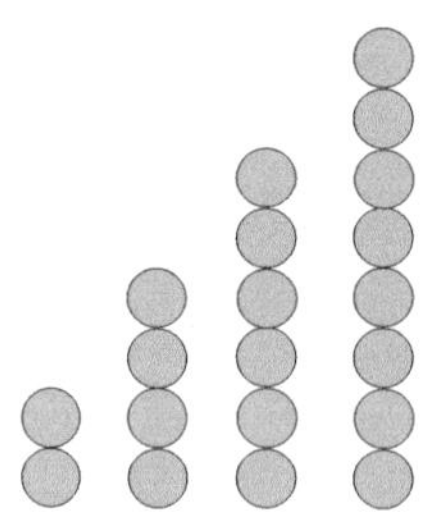	Use connecting cubes to show a growing pattern that begins with zero. Ask students to write this pattern with numbers on their dry erase boards. Explain how growing patterns can begin with any amount as long as they grow by a constant amount. Demonstrate a "grow-by-two" pattern that begins with one. Encourage students to describe and continue the pattern. Ask students to write the pattern (1, 3, 5, 7, . . .). Provide students time to create, describe, and write other complex growing patterns.
Questions to Assess	♦ What comes next in the growing pattern? ♦ How do you know?	♦ How would you describe the pattern? ♦ How are the towers different?	♦ What is next in your pattern? ♦ How many counting chips will be used in the sixth term?	♦ How could you record this pattern using numbers? ♦ What is the next term? How do you know?

Traditional Formative Assessment

True or False. Write T or F.

$2 + 3 = 3 + 2$	$4 + 1 = 5 + 1$	$2 + 1 + 3 = 1 + 2 + 3$
$4 - 2 = 2 - 4$	$6 + 3 = 3 + 6$	$10 - 1 = 1 - 10$

Limitations: The problem with potential "lucky guess" answers is exacerbated with True/False questions.

Enhanced Formative Assessment

Note: This assessment can be modeled with objects or completed on paper. Record students' descriptions.

$$7 + 3 = 3 + 7$$

Is this correct? Explain your thinking.

Ivan

$7 + 3 = 3 + 7$

Is this correct? Explain your thinking.

No Wrug

Ivan does not yet understand commutativity. He needs to work with comparing equal quantities that are composed in different ways.

Instructional Focus: **Rebuild**

Jocelyn

$7 + 3 = 3 + 7$ 9

Is this correct? Explain your thinking.

Jocelyn does not yet understand commutativity. She will benefit from learning how to accurately compose 10 in different ways.

Instructional Focus: **Core 1**

Kevin

√ $7 + 3 = 3 + 7$

Is this correct? Explain your thinking.

YES BUT FIX

Kevin understands something about commutativity, but does not yet have full understanding because he thinks it needs to be fixed.

Instructional Focus: **Core 2**

La'Nya

$7 + 3 = 3 + 7$ order does't matter

Is this correct? Explain your thinking.

yes 10 = 10

La'Nya understands commutativity and even explains that the order of the addends is irrelevant to the sum. She further proves her understanding of commutativity by including the equation 10 = 10.

Instructional Focus: **Challenge**

CCSS Algebraic Thinking: Commutativity

	Rebuild Focus	Core 1 Focus	Core 2 Focus	Challenge Focus
	Red and Yellow	**Show Ten**	**Flip Facts**	**Triple Addends**
Goal	Composing basic quantities in different ways	Composing 10 in different ways	Recognizing that the order of the addends does not change the sum	Understanding commutativity with several addends
Materials	♦ Bi-colored chips (yellow and red)	♦ Ten frames ♦ Counting chips	♦ Connecting cubes ♦ Dry erase board and markers	♦ Three dice (each labeled 1–6) ♦ Box lid ♦ Dry erase boards and markers
Suggested Activity Directions	Show 3 chips on the red side and compare these to 3 chips shown on the yellow side. Line chips up to show how each red-sided chip matches each yellow-sided chip. Discuss how the quantities are the same: 3 = 3. Flip the counters to show different combinations of three while keeping them aligned (2 + 1 = 1 + 2). Repeat with a quantity of 4 chips, then 5 chips. Encourage students to model with their own chips. If appropriate, try greater quantities.	Give each student a 10-frame and some chips (2 colors). Ask students to show 10 using 5 chips of one color and 5 of a different color (5 + 5). Encourage students to show 10 another way such as 6 and 4. Ask some students to show 6 + 4 and some students show 4 + 6. Discuss how the parts are equal but presented in a different order. Encourage students to name the parts (e.g., 6 blue and 4 green is the same as 4 blue and 6 green). Repeat with other ways to compose 10 (7 + 3, 8 + 2, 9 +1).	Connect a row of cubes by placing 7 blue cubes on the left and 3 yellow cubes on the right. Ask students to write the expression (7 + 3) on their dry erase boards. Flip the row of cubes and have students write the expression (3 + 7). Show students how to connect expressions into an equation (7 + 3 = 3 + 7). Help students understand commutativity by flipping the row back and forth while students say the expressions. Try other combinations and quantities if appropriate.	Toss all three dice into the box lid. Ask students to write expressions using the three numbers as addends. Provide time for students to compare various ways the expression can be recorded. Demonstrate how to write two of the expressions in the form of an equation (e.g., 3 + 1 + 4 = 1 + 4 + 3). Toss the dice again. Ask students to use the numbers to write two different expressions in the form of an equation (e.g., 5 + 2 + 3 = 2 + 5 + 3). Continue. Try four addends, if appropriate.
Questions to Assess	♦ How do you know that the quantities are the same? ♦ Is there another way to compose this number?	♦ How is 10 composed? ♦ What is the total?	♦ How can you record the expression? ♦ What happens if you change the order of the addends?	♦ Does the order of the addends change the sum? ♦ What if you were to include subtraction?

Traditional Formative Assessment

Circle the expressions for 18.

3 + 5 + 11	12 + 2 + 6 + 1	4 + 3 + 3 + 4 + 4
4 + 4 + 3 + 3	4 + 4 + 3 + 3 + 2 + 2	8 + 3 + 2 + 1 + 1 + 2

Limitations: Students can calculate rather than use problem solving and reasoning to obtain the correct answers.

Enhanced Formative Assessment

The stools have 3 legs and the tables have 4 legs.

If there are 18 legs, how many stools and how many tables are there?

Maria

The stools have 3 legs and the tables have 4 legs.

If there are 18 legs, how many stools and how many tables are there?

3 + 4 = 7

Maria does not use problem solving or numeric reasoning. She takes the two numbers given in the problem and adds them together.

Instructional Focus: **Rebuild**

Nathaniel

The stools have 3 legs and the tables have 4 legs.

If there are 18 legs, how many stools and how many tables are there?

3 + 3 + 4 + 4

Even though Nathaniel's answer is incorrect, he understands that he needs to combine groups of 3 and groups of 4 to solve this problem.

Instructional Focus: **Core 1**

Omar

The stools have 3 legs and the tables have 4 legs.

If there are 18 legs, how many stools and how many tables are there?

||| |/\ |\\| |||| \|||

2 stools 3 tables

Omar correctly solves the problem and uses pictures (tally marks of legs) to explain how he obtained his answer.

Instructional Focus: **Core 2**

Paula

The stools have 3 legs and the tables have 4 legs.

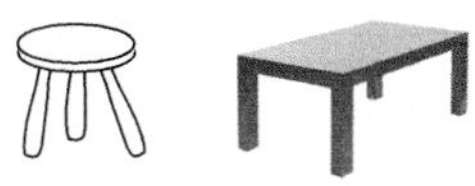

If there are 18 legs, how many stools and how many tables are there?

t(4)	S(3)	L
3	2	18
0	6	18

Paula uses numeric reasoning to accurately solve this problem in two ways. She presents her answers in the form of a data table.

Instructional Focus: **Challenge**

	Rebuild Focus	Core 1 Focus	Core 2 Focus	Challenge Focus
	Buttons and Holes	**Stools and Tables**	**Spiders and Ants**	**Wheels and Wheels**
Goal	Understand the problem	Represent the problem	Search for more than one solution	Solve more complex problems
Materials	♦ Highlighters ♦ Red Buttons (2 holes) ♦ Blue Buttons (4 holes)	♦ Popsicle sticks ♦ Dry erase boards and markers	♦ Pipe cleaner pieces ♦ Index cards and markers	♦ Chart paper and markers
Suggested Activity Directions	Present this problem; The red buttons have 2 holes and the blue buttons have 4 holes. If there are 8 holes, how many red and blue buttons? Ask: What do we know? Ask students to identify (discuss and highlight) the important information in the number story. Ask: What do we need to know? Help students restate the number story question, "How many red buttons and blue buttons if there are 8 holes?" Encourage students to count the holes in the buttons to find solutions. Try a different number of button holes if appropriate. 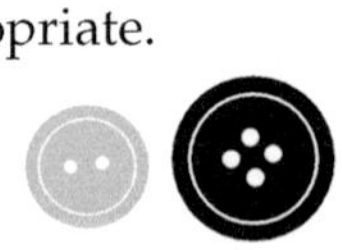	Use the stools and tables problem. Help students model the problem by making groups of legs with popsicle sticks. When students have modeled both solutions to the problem, provide time for them to draw a representation of each solution on their dry erase boards. Try the same scenario but change the number of legs. Encourage students to represent the solutions with popsicle sticks and with pictures. If appropriate, have students record the matching equations.  	Present this problem: Spiders have 8 legs and ants have 6 legs. What are all the possible combinations of spiders and ants if there are 48 legs in the bug jar? Encourage students to use pipe cleaner pieces to group legs. Provide time for students to write different solutions on separate index cards. Help students recognize repeated solutions and how to organize the information to show all three possible solutions. 6 Spiders 0 Ants 3 Spiders 4 Ants 0 Spiders 8 Ants	Present this problem: Bicycles have 2 wheels, tricycles have 3 wheels, and wagons have 4 wheels. If there are 24 wheels, how many bicycles, tricycles, and wagons could there be? Ask students to create an organized table to show all of the possible solutions. W | T | B 6 | 0 | 0 5 | 0 | 2 4 | 2 | 1 4 | 0 | 4 3 | 4 | 0 3 | 2 | 3 3 | 0 | 6 2 | 4 | 2 2 | 2 | 5 2 | 0 | 8 1 | 6 | 1 1 | 4 | 4 0 | 8 | 0 0 | 6 | 3 0 | 4 | 6 0 | 2 | 9 0 | 0 | 12
Questions to Assess	♦ What do we know? ♦ What do we need to find out?	♦ How can you show your thinking using the manipulatives? ♦ Does the picture match the model?	♦ How many solutions can you find? ♦ Is the solution the same or different?	♦ How can you keep the information organized? ♦ Are there any other solutions?

Traditional Formative Assessment

The plant grows 3 centimeters every day. How tall will the plant be on day 7?

- O 10 centimeters
- O 12 centimeters
- O 20 centimeters
- O 21 centimeters

Limitations: As with other multiple choice questions, the student may not know the answer, but simply have a lucky guess.

Enhanced Formative Assessment

The Function Machine always follows the same rule.
Complete the chart and name the rule.

In	Out
2	5
5	8
8	11
11	

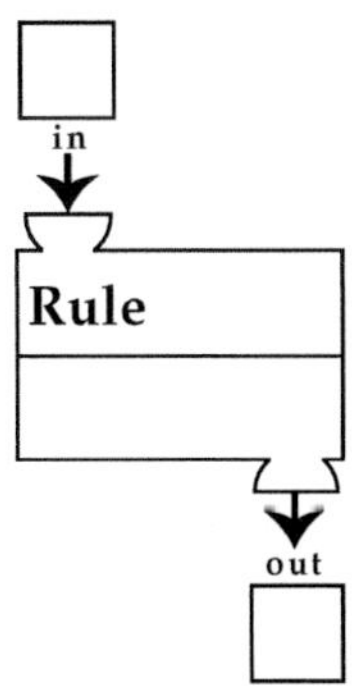

Quincy

The Function Machine always follows the same rule.

Complete the chart and name the rule. 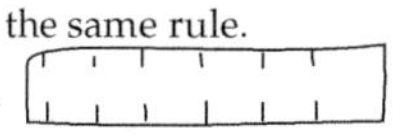

In	Out
2	5
5	8
8	11
11	5
2	8
5	11
8	5

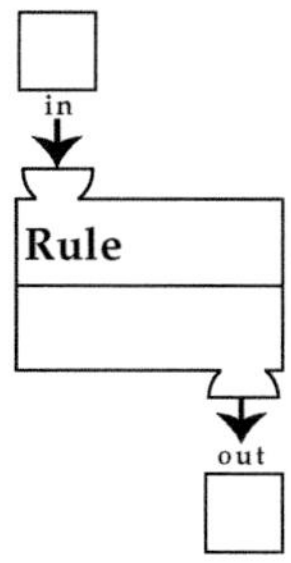

Quincy draws a ruler to show the "rule" and creates repeating patterns. He needs to work with more simplistic concepts of change.

Instructional Focus: **Rebuild**

Raheem

The Function Machine always follows the same rule.

Complete the chart and name the rule. one more

In	Out
2	5
5	8
8	11
11	12
12	13
13	14

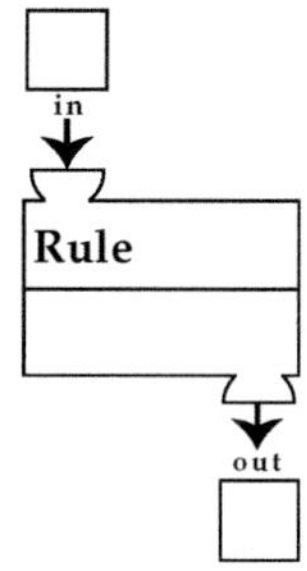

Raheem has some understanding of functions, but does not recognize the specific change presented on the chart.

Instructional Focus: **Core 1**

Sophie

The Function Machine always follows the same rule.

Complete the chart and name the rule. Add 2

In	Out
2	5
5	8
8	11
11	13
13	16
16	18

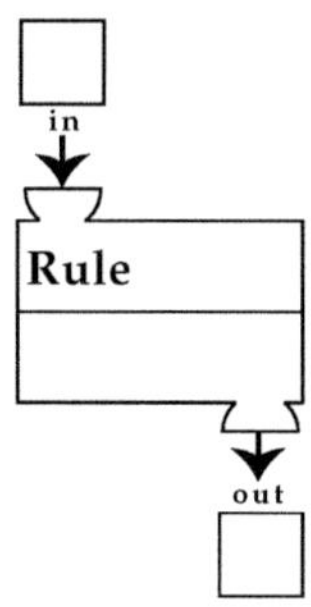

Sophia recognizes concepts of change, but makes a computational error and mistakenly identifies the change as "add 2."

Instructional Focus: **Core 2**

Trinity

The Function Machine always follows the same rule.

Complete the chart and name the rule.

In	Out
2	5
5	8
8	11
11	14
14	17
17	20

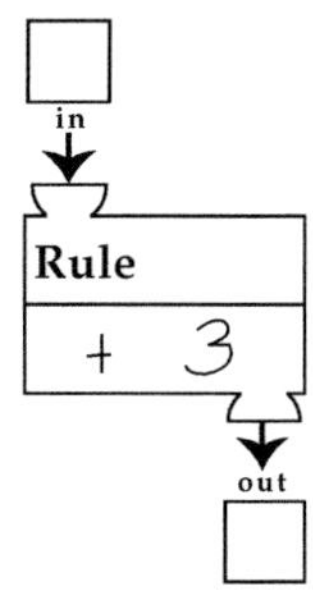

Trinity accurately completes the chart and identifies the changes as "+3."

Instructional Focus: **Challenge**

Algebraic Thinking: Concepts of Change

<table>
<tr><th></th><th>Rebuild Focus</th><th>Core 1 Focus</th><th>Core 2 Focus</th><th>Challenge Focus</th></tr>
<tr><th></th><th>Change It!</th><th>Follow the Leader</th><th>What's the Rule?</th><th>Function Junction</th></tr>
<tr><th>Goal</th><td>Recognize the "plus 1" concept of change as a rule</td><td>Recognize the "plus 2" concept of change</td><td>Identify and compute the +2 and +3 concepts of change</td><td>Recognize and apply more complex concepts of change</td></tr>
<tr><th>Materials</th><td>♦ Number lines (0–20)
♦ Dry erase boards and markers</td><td>♦ Connecting cubes</td><td>♦ Laminated In/Out charts, markers, and erasers</td><td>♦ Laminated In/Out charts, markers, and erasers</td></tr>
<tr><th>Suggested Activity Directions</th><td>Distribute number lines to students. Encourage students to discuss how the numbers are in order and increase by 1. Using this experience as background knowledge, write the number 1 on a dry erase board. Invite students to apply the plus 1 rule. Say: Use the plus 1 rule to change it. Encourage students to write 2 on individual dry erase boards. Repeat many times using different numbers (2–10) in random order. If students are ready, try numbers 11–20.
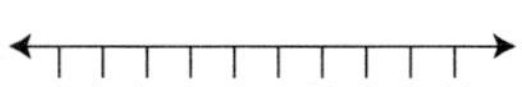
</td><td>Serve as the "leader" by showing the students a tower of 2 connecting cubes. Ask each student to follow the leader by using connecting cubes to make a tower that shows the plus 2 concept of change. Now show a tower of 3 cubes and ask the students to show the plus 2 concept of change. Repeat with various towers of cubes. Highlight the idea that students are becoming a function machine for the +2 rule. Provide opportunities for students to be the leader.</td><td>Plug in even numbers that show a "plus 2" function on your chart. Ask students to copy the information on their charts. Discuss the concept of change presented. Erase and ask students to create a chart that shows the "plus 2" rule, but uses odd numbers. Provide time for students to share their function charts and verify the rule. Erase and make a different chart that shows the "plus 3" concept of change. Discuss the pattern of even and odd numbers. Try other +3 function charts.</td><td>Ask students to create various function charts (+4, +6, +7). Provide time for students to analyze one another's charts and name the rule. Introduce subtraction as a concept of change. Encourage students to create subtraction function charts to analyze. If students are ready, challenge them by demonstrating a multiplication concept of change. Invite students to try to create their own.<table><tr><th>In</th><th>Out</th></tr><tr><td>3</td><td>6</td></tr><tr><td>4</td><td>8</td></tr><tr><td>5</td><td>10</td></tr></table></td></tr>
<tr><th>Questions to Assess</th><td>♦ What if we change it by the plus 1 rule?
♦ Why do you think it is called a rule?</td><td>♦ What is the function?
♦ How can you show the rule?</td><td>♦ What do you notice about the numbers?
♦ How can you prove the rule?</td><td>♦ How do you know the rule?
♦ How could you make a more complex rule?</td></tr>
</table>

Traditional Formative Assessment

Solve.

$12 + n = 9 + 6$ $\qquad$ $7 + 2 = 2y + 1$

$n = ?$ $\qquad$ $? = y$

Limitations: The representation of unknowns (letters) may confuse or intimidate primary students. A different way to reveal the student's level of algebraic thinking is needed.

Enhanced Formative Assessment

It is all about balance.

How much is 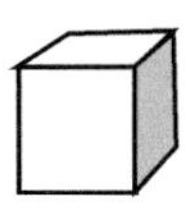 ?

How much is 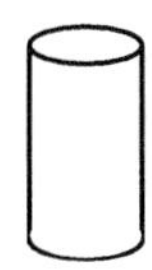 ?

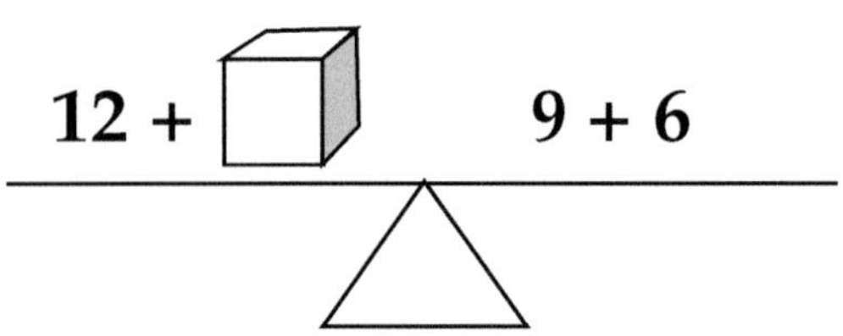

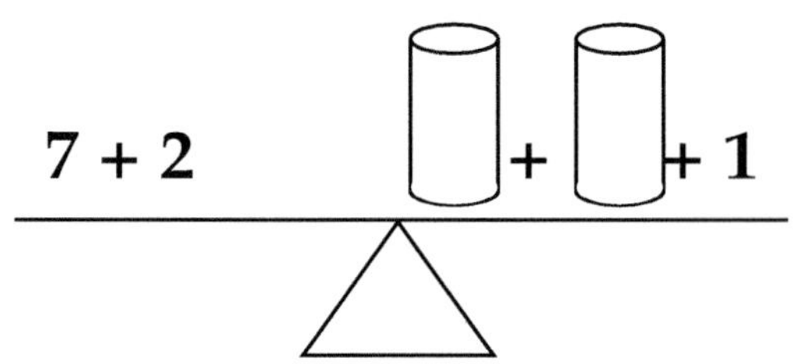

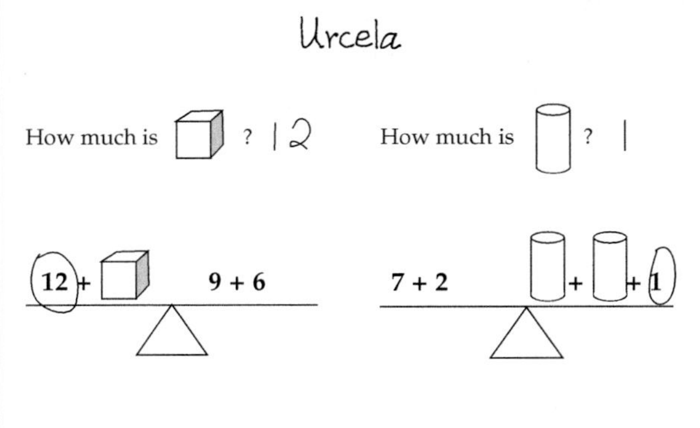

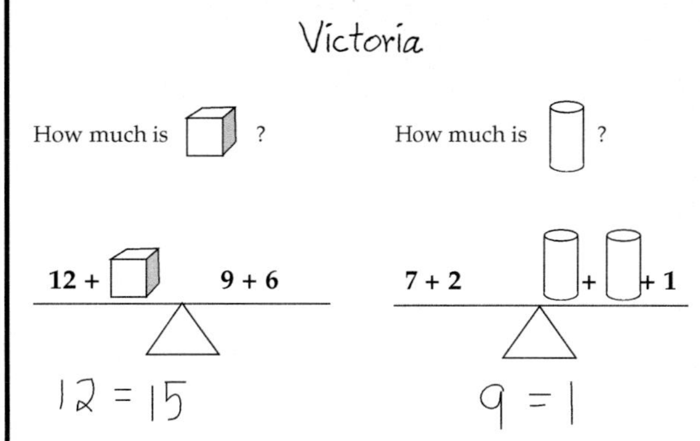

Urcela inaccurately assigns the unknown numerical variables the values that are represented on the same side of the balanced equation.

Instructional Focus: **Rebuild**

Victoria knows that she needs to combine the addends, but she does not understand how the unknown numerical variables fit into the equations. She creates inequalities.

Instructional Focus: **Core 1**

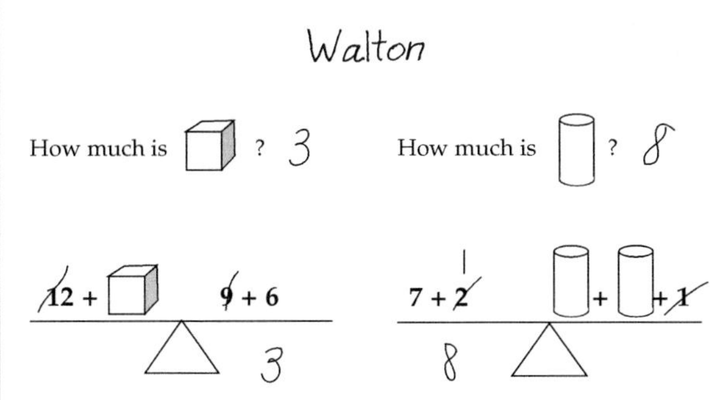

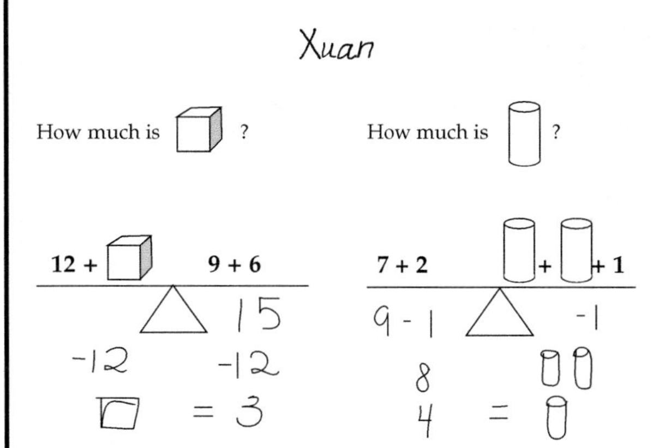

Walton has a relatively strong understanding of how to use balance and equality to solve for unknown numerical variables. However, he does not know how to address multiple unknowns.

Instructional Focus: **Core 2**

Xuan clearly understands how to use balance and equality to solve for unknown numerical variables. He accurately subtracts to isolate the unknown numerical variable.

Instructional Focus: **Challenge**

CCSS Algebraic Thinking: Algebraic Symbols, Equality, & Balance

	Rebuild Focus	Core 1 Focus	Core 2 Focus	Challenge Focus
	Part-Part-Whole	**Secret Cup**	**More Secret Cups**	**Keep the Balance**
Goal	Understanding missing addends	Connecting missing addends and unknown numerical variables	Solving multiple unknown variables in addition equations	Subtracting unknown variables
Materials	♦ Two dice (numerals) ♦ Counters ♦ Chart paper and markers ♦ Sticky notes	♦ Pieces of cardboard (approx. 12″ × 2″) ♦ Numeral tiles (0–9, +) ♦ Small cups	♦ Pieces of cardboard (approx. 12″ × 2″) ♦ Several sets of numeral cards (0–25) ♦ Small cups	♦ Laminated sentence strip and dry erase markers ♦ Counting chips
Suggested Activity Directions	Roll the dice. The numerals represent addends. Students show the addends and find the sum using counters. Record the equation on chart paper (e.g., 2 + 3 = 5). Repeat until at least four equations have been recorded. Ask students to close their eyes while you place sticky notes on one addend in each equation. Encourage students to use the sum and the known addend to figure out the hidden or missing addend. Use counters to model if necessary.	Use a piece of cardboard to represent the arm on a balance. Hold it in the middle and prompt discussion about balance and equality. Use digit tiles to show 4 + 3 on one side and ask which tile/s could be placed on the other side to maintain balance. Ask students to create other equalities. Show 7 + 3 on one side and ask students to close their eyes while you place 6 + 4 on the other side. Cover one of the numbers with a cup. Ask students to open their eyes and name the unknown. Repeat with other equations.	Using the same ideas as in the Core 1 Focus, discuss balance and equality. Use the numeral cards to create balanced equations. Use a cup to make one of the addends unknown. Extend the activity by using more than one cup. Note: The cups always cover the same value within a specific equation. For example, 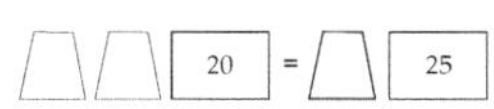 Ask students to create expressions with unknowns covered with cups. Encourage students to solve one another's unknown variables.	Write this equation on the sentence strip using the same color chips for the unknowns: 4 + 1 + ●= 3 +● + ● Ask students how they can determine the unknown numeric variables. Suggest removing one chip from each side. Ask: Will the equation still be balanced? (Yes). Remove a chip from each side and help students solve, 5 = 3 + ● Provide time for students to write and solve the following equations; 3 + ● + ● = ● + 7 10 + 2 + ● = 9 + ● + ●
Questions to Assess	♦ How can we use the sum to figure out the missing addend? ♦ Which part is hiding?	♦ What value is under the cup? ♦ How can you solve for the unknown?	♦ How can you use addition and subtraction to solve for the unknown variable?	♦ How can you maintain balance? ♦ Which values are known?

4

Geometry

shapes

direction location

flips

symmetry turns slides two-dimensional

three-dimensional distance

coordinates

CCSS Geometry: Two-Dimensional Shapes

Traditional Formative Assessment

Name this shape.

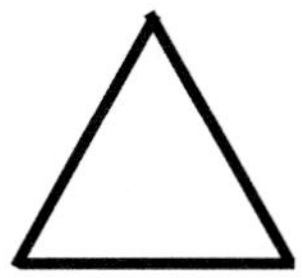

- O circle
- O triangle
- O rectangle

Limitations: In addition to the flaws associated with multiple choice questions, a young student may know the shape name, but not be able to read the shape name.

Enhanced Formative Assessment

Say the shape names and ask students to draw or place (from a set of cutout shapes) each shape in the correct box.

Circle	Triangle	Rectangle

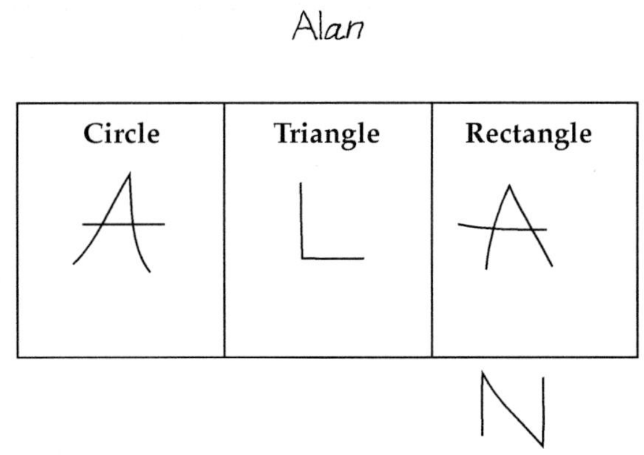

Alan has some understanding of one-to-one correspondence because he places one letter of his name in each box, but he does not appear to have any knowledge of two-dimensional shapes.

Instructional Focus: **Rebuild**

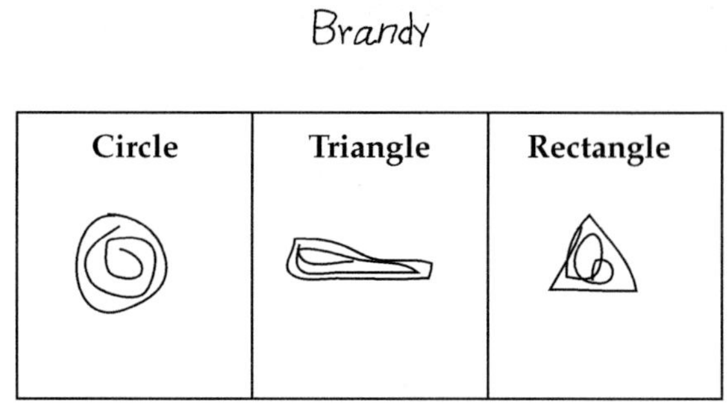

Brandy accurately sketches a circle, but mixes up the triangle and rectangle.

Instructional Focus: **Core 1**

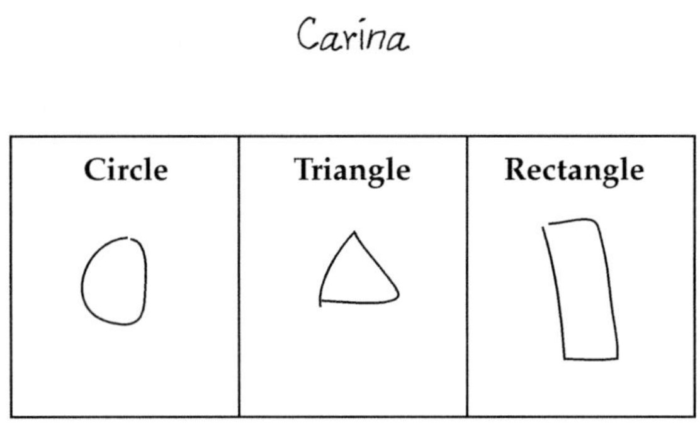

Carina correctly sketches all three two-dimensional shapes. She is ready to learn about other types of rectangles (squares).

Instructional Focus: **Core 2**

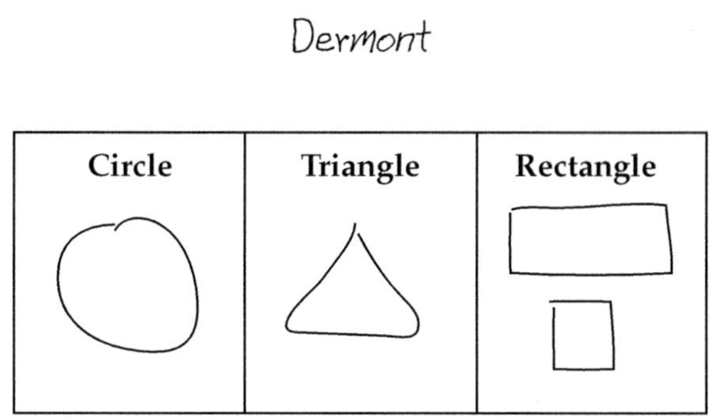

Dermont has a high level of understanding of basic two-dimensional shapes. He even understands that squares are a subset of the rectangle group.

Instructional Focus: **Challenge**

Geometry: Two-Dimensional Shapes

<table>
<tr><th></th><th>Rebuild Focus</th><th>Core 1 Focus</th><th>Core 2 Focus</th><th>Challenge Focus</th></tr>
<tr><th>Goal</th><td>Sort triangles and two-dimensional shapes that are not triangles</td><td>Compare triangles, rectangles, and circles</td><td>Describe square rectangles and non-square rectangles</td><td>Identify and describe more complex polygons</td></tr>
<tr><th>Materials</th><td>♦ Cutout shapes (including different types of triangles)
♦ Poster paper and markers</td><td>♦ Attribute blocks
♦ Paper bag</td><td>♦ Geoboards and Geobands (rubber bands)</td><td>♦ Geoboards and Geobands (rubber bands)</td></tr>
<tr><th rowspan="2">Suggested Activity Directions</th><th>Triangle or Not?</th><th>What's in the Bag?</th><th>Square Rectangles</th><th>More Polygons</th></tr>
<tr><td>Draw a large triangle on the poster paper. Distribute shape cutouts to students. Ask students to take turns placing triangles inside the triangle on the poster and "not triangles" outside of the triangle on the poster. Encourage students to explain why (give attributes of the two-dimensional face on the cutout shapes) as they place the shapes in the correct location. If appropriate, try the same activity with other shapes.
</td><td>Prompt discussion about the two-dimensional shapes represented by the attribute blocks. Place blocks in a bag. Say: "I will reach in the bag and find a triangle without looking." Use a think-aloud strategy, such as explaining that you are feeling for a shape that has a face with 3 straight sides. Students take turns looking for rectangles, circles, or triangles.
</td><td>Distribute materials and ask students to use one geoband to make a rectangle on their geoboards. Compare rectangles, noting size and position. Discuss the commonalities (polygon with 4 straight sides and 4 right angles). Explain to students that these same characteristics apply to a square; therefore, squares are part of the rectangle group. Provide time for students to take turns creating and sharing square and nonsquare rectangles.
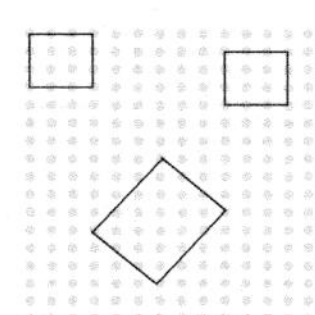</td><td>Share with students that polygons are closed shapes with all straight sides. Allow students to make various polygons (regular and irregular) with geobands on geoboards. Provide time for students to report the number of sides the shape has and provide the name of the shape for the students.<table><tr><th>Sides</th><th>Name</th></tr><tr><td>5</td><td>pentagon</td></tr><tr><td>6</td><td>hexagon</td></tr><tr><td>7</td><td>heptagon</td></tr><tr><td>8</td><td>octagon</td></tr><tr><td>9</td><td>nonagon</td></tr><tr><td>10</td><td>decagon</td></tr></table></td></tr>
<tr><th>Questions to Assess</th><td>♦ Is it a triangle?
♦ How do you know it is not a triangle?</td><td>♦ What shape do you feel?
♦ How do you know it is a rectangle?</td><td>♦ Why is a square part of the rectangle group?
♦ How many corners does a square have?</td><td>♦ How many sides does your polygon have?
♦ Are all of the sides the same size?</td></tr>
</table>

Traditional Formative Assessment

Name this shape.

- ◯ circle
- ◯ sphere
- ◯ cylinder
- ◯ rectangle
- ◯ cube

Limitations: Two-dimensional drawings of three-dimensional shapes can be very confusing for students even if they have knowledge of the shape names.

Enhanced Formative Assessment

Place these three-dimensional shapes in front of students (cube, cylinder, cone, sphere).

Say: "Point to the cylinder." If student is correct, ask: "How would you describe it?"

Say: "Point to the cone." If student is correct, ask: "How would you describe it?"

Say: "Point to the sphere." If student is correct, ask: "How would you describe it?"

Say: "Point to the cube." If student is correct, ask: "How would you describe it?"

CCSS Geometry: Three-Dimensional Shapes

Elsie

cylinder	✗	
cone	✗	
sphere	✗	
cube	✗	

Elsie is unable to identify any of the three-dimensional shapes.

Instructional Focus: **Rebuild**

Fabio

cylinder	✓	It is tall.
cone	✗	
sphere	✗	
cube	✓	It is not tall.

Fabio identifies two of the shapes. His descriptions are not applicable to all cylinders and cubes.

Instructional Focus: **Core 1**

Geno

cylinder	✓	It is like a can.
cone	✓	It is like ice cream.
sphere	✓	It is like a ball.
cube	✓	It is like a box.

Geno accurately identifies all four three-dimensional shapes. His descriptions are simplistic and lack geometry vocabulary.

Instructional Focus: **Core 2**

Harmony

cylinder	✓	Two circles and a curved rectangel
cone	✓	One circle and a point on top.
sphere	✓	All curved. No edges or points.
cube	✓	A lot of edges and six squares.

Harmony accurately identifies all four three-dimensional shapes. Her descriptions include rich math vocabulary.

Instructional Focus: **Challenge**

CCSS Geometry: Three-Dimensional Shapes

	Rebuild Focus	Core 1 Focus	Core 2 Focus	Challenge Focus
	Cylinder Search	**Cones and Spheres**	**Let's Make Nets**	**Pyramids and Prisms**
Goal	Identify and describe cylinders	Identify and sort cones and spheres	Make nets of cubes and cylinders	Identify and describe more complex three-dimensional shapes
Materials	♦ Cylinders of various sizes ♦ Other three-dimensional shapes	♦ Cubes and spheres of various sizes	♦ Cubes and cylinders of various sizes ♦ Scissors, tape ♦ Graph paper, pencils	♦ Pyramids and prisms of various sizes ♦ Pipe cleaners
Suggested Activity Directions	Display the shapes. Choose a cylinder and hold it up for students to see. Describe it (curved surface and one circle at each end). Ask students to find other cylinders in the group of shapes. Each student should find and describe at least one cylinder. Go on a cylinder search around the classroom. Encourage students to describe the cylinders they find. 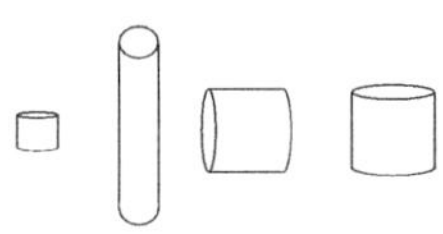	Display the set of cones and spheres. Ask students to describe the shapes. Ask: "Which shapes do you think are cones?" Ask students to explain their answers. Provide vocabulary as needed (point, flat, circle, curved surface). Focus students' attention on the spheres. Ask: "Do you think spheres can roll? Why?" Encourage students to name the differences between spheres and cones. 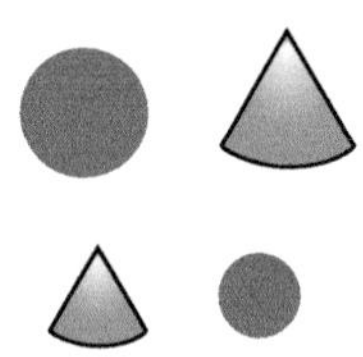	Ask students to trace the cube on graph paper. Demonstrate how to move the cube so that all 6 faces can be traced with adjoining edges. Encourage students to trace the faces and cut out the net. Ask students to fold on adjoining face edges and tape other edges to construct the cube. To create the net of a cylinder, place curved edge on graph paper and roll one full rotation. Draw the area and then trace the circle on each end. Cut out and tape together. If students are ready, try constructing the net of a cone. 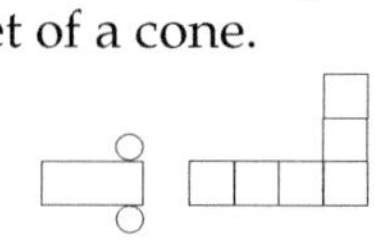	Help students sort the shapes. Encourage students to distinguish between shapes with triangular faces and shapes with rectangular faces. Note: Pyramids have triangular faces and a base that identifies the type of pyramid (triangular pyramid, square pyramid, pentagonal pyramid, hexagonal pyramid, etc.). Prisms have rectangular faces and two bases that identify the type of prism (e.g., triangular prism, rectangular prism, pentagonal prism, hexagonal prism). Encourage students to construct the shapes using pipe cleaners to represent the edges.
Questions to Assess	♦ Is it a cylinder? How do you know? ♦ How short can a cylinder be?	♦ What is the difference between a cone and a sphere? ♦ How could you describe the shape?	♦ How will you put the net together? ♦ How is the shape made?	♦ How many faces on the shape? ♦ How do you know that you have all the faces for the net?

Traditional Formative Assessment

Match the correct answers.

Location means ___________.	which way
Direction means ___________.	how far
Distance means ___________.	where

Limitations: Knowledge of the definition does not mean the student understands the concept. Likewise, a student may understand the concept and not know the definition.

Enhanced Formative Assessment

Where is the frog?
Which way should the frog hop to the circle?
How far is the star?

CCSS Geometry: Location, Direction, and Distance

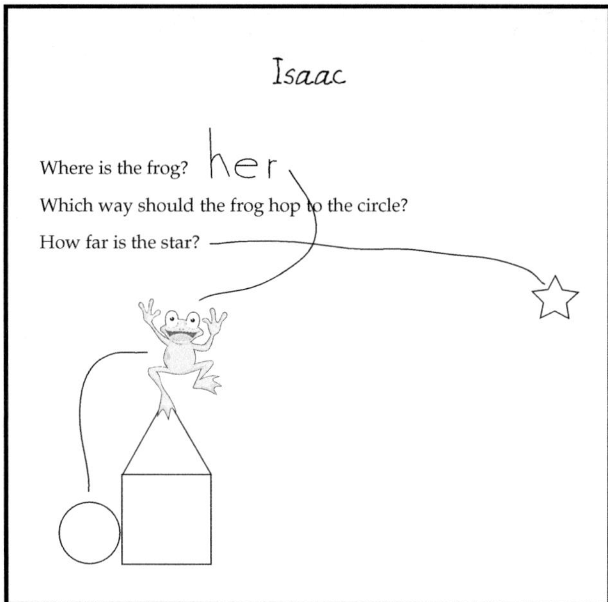

Isaac draws lines to show location, direction, and distance. He writes "here" to describe the location of the frog. He does not use any geometry vocabulary.

Instructional Focus: **Rebuild**

Josie writes that the frog is located on the "roof top." She uses simplistic geometry vocabulary to indicate direction and distance (up and down).

Instructional Focus: **Core 1**

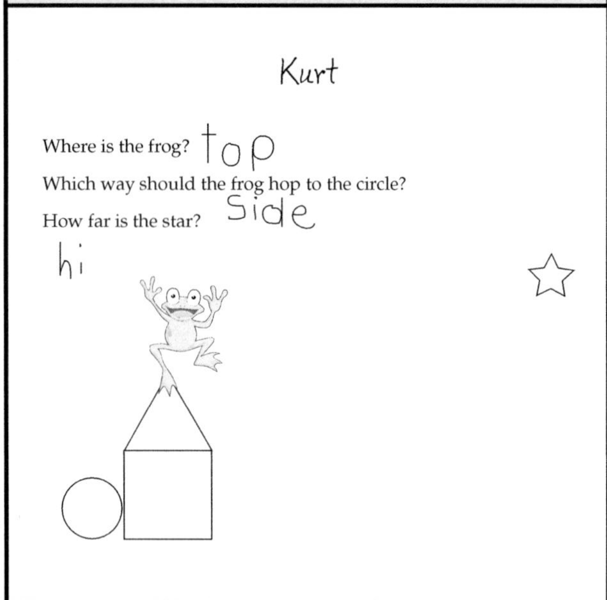

Kurt's descriptions of the location, direction, and distance are accurate but lack details.

Instructional Focus: **Core 2**

Lisha uses a high level of geometry vocabulary to describe location and direction. Her description of the distance (farther) lacks a comparison (farther than what?).

Instructional Focus: **Challenge**

CCSS Geometry: Location, Direction, and Distance

	Rebuild Focus	Core 1 Focus	Core 2 Focus	Challenge Focus
	Stack the Blocks	**Bean Bag Toss**	**Moving Stick Figures**	**Spill the Beans**
Goal	Understand simple locations such as top, middle, and bottom	Describe location and distance using near, far, and between	Understand more complex directions (right and left)	Describe distance using comparative (er) and superlative (est)
Materials	♦ Colored blocks (red, blue, yellow)	♦ Bean bags (different colors) ♦ Box lid	♦ Popsicle sticks ♦ Paper circles ♦ Tape ♦ Pencils	♦ Uncooked beans (variety of colors and sizes) ♦ Cups ♦ Pieces of paper
Suggested Activity Directions	Give each student a red block and a blue block. Give directions: Place the red block on top of the blue block. Ask: "Where is the blue block?" After students respond, switch the blocks. Ask students to describe the location of the blocks using the words, top and bottom. Introduce the word, middle, using the yellow block. Change how the blocks are stacked and ask student to describe the locations using the words, top, middle, and bottom. 	Draw a star in the center of the box lid. Students take turns tossing bean bags into the box lid. After each round of tosses, discuss the placement of the bean bags using the words, near, far, and between to describe the locations and distances (e.g., The red bean bag is near the star. The green bean bag is between the blue and orange bean bags.). 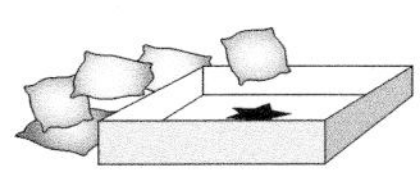	Ask students to draw faces on the circles with right hands and with left hands. Discuss hand preferences. Are you right handed or left handed? Tape a face to a popsicle stick for each student. Say: "Hold your stick figure with your right hand and move it slowly to the right. Switch hands and move the stick figure quickly to the left." Give other directions such as: hop to the right; swim to the left; run to the right; dance to the left. Add up and down to the right or left directions. 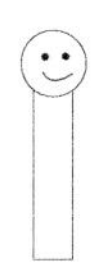	Give each student a small handful of beans in a cup. Students take turns spilling the beans on paper. After each spill, discuss the comparative and superlative distances. For example: This red bean is closer to the center of the paper. That white bean is furthest from the center of the paper. Continue spilling beans and describing distances using the words near, close, closer, closest, far, farther, and farthest. 
Questions to Assess	♦ Which block is on top? ♦ Where is the yellow block?	♦ Is it far from the star? ♦ Which bag is between the others?	♦ Which way is left? How do you know? ♦ What if you move the figure down to the left?	♦ Which bean is farthest from the top of the paper? ♦ Which bean is closer to the bottom of the paper?

Geometry: Basic Coordinates

Traditional Formative Assessment

Whatarethecoordinatesofthetriangle?

- O 3,3
- O 3,4
- O 4,3
- O 4,4

What are the coordinates of the circle?

- O 2,2
- O 2,5
- O 5,2
- O 5,5

Coordinate Grid

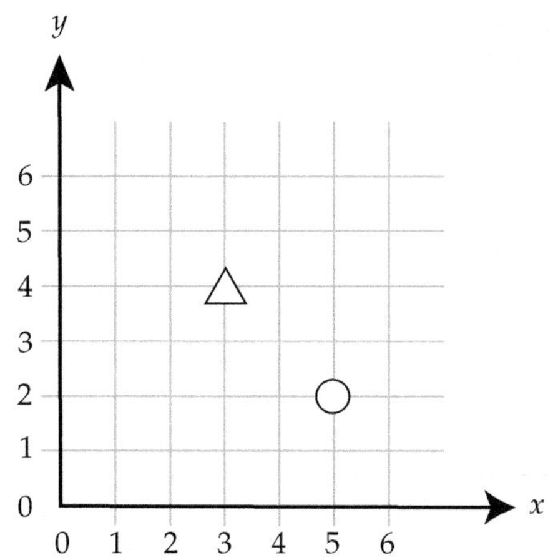

Limitations: As with other multiple choice questions, the student may not know the answer but simply have a lucky guess.

Enhanced Formative Assessment

Let's work with coordinates!

Describe the location of the triangle.

Describe the location of the circle.

Coordinate Grid

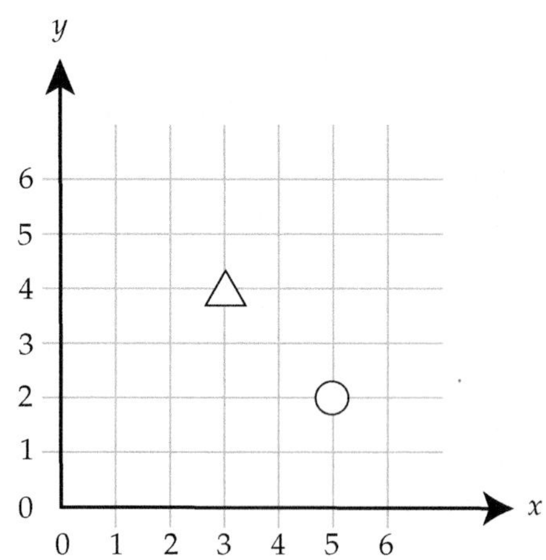

Geometry: Basic Coordinates

Mia

Let's work with coordinates!

Describe the location of the triangle.
on graft paper

Describe the location of the circle.
on graft paper

Coordinate Grid

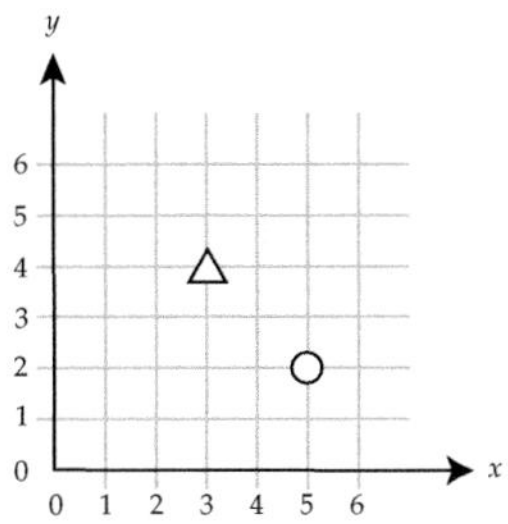

Mia describes the location of the shapes as on the graph paper. She does not use the coordinates.
Instructional Focus: **Rebuild**

Nelson

Let's work with coordinates!

Describe the location of the triangle. up 4

Describe the location of the circle. down 2

Coordinate Grid

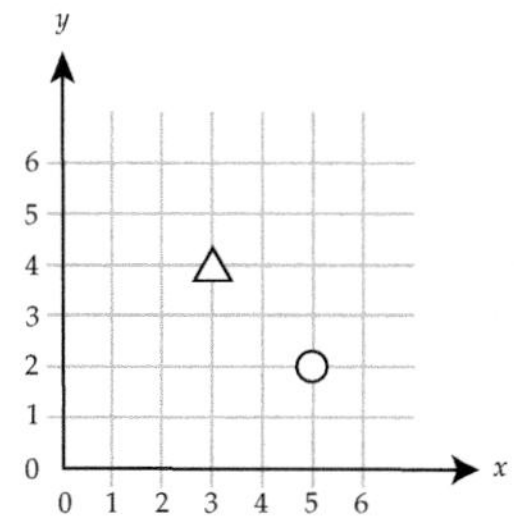

Nelson notices the y coordinates but does not address the x coordinates. He includes directional vocabulary (up, down).
Instructional Focus: **Core 1**

Orlando

Let's work with coordinates!

Describe the location of the triangle.
4 + 3

Describe the location of the circle.
2 + 5

Coordinate Grid

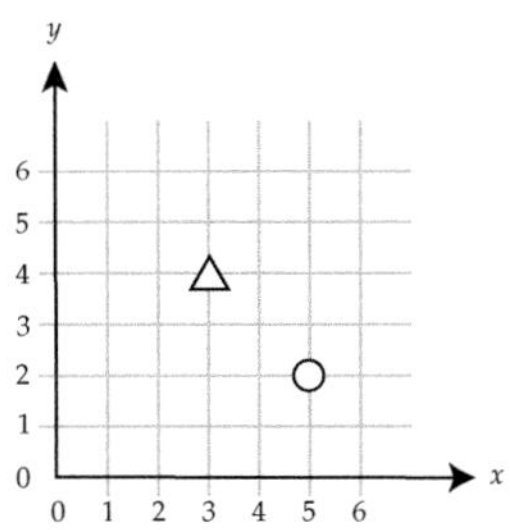

Orlando addresses both coordinates but does not know how to communicate the coordinates in written form. He erroneously creates a numeric expression for each location.
Instructional Focus: **Core 2**

Portia

Let's work with coordinates!

Describe the location of the triangle.
3, 4

Describe the location of the circle.
5, 2

Coordinate Grid

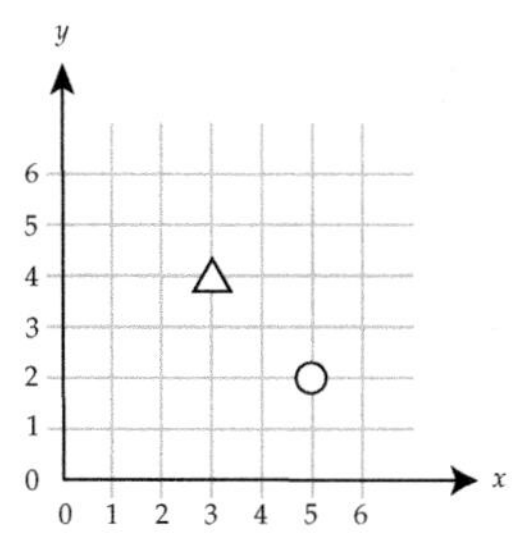

Portia understands how to accurately describe the location of the shapes with coordinates.
Instructional Focus: **Challenge**

Geometry: Basic Coordinates

	Rebuild Focus	Core 1 Focus	Core 2 Focus	Challenge Focus
	Make a Path	**Crazy Coordinates**	**Fast Coordinates**	**Complex Coordinates**
Goal	Mark paths on a coordinate grid	Understand *x* and *y* coordinates using numbers and letters	Communicate coordinates using commas	Explore negative coordinates
Materials	♦ Graph paper (inch) ♦ Stickers ♦ Markers	♦ Graph paper (inch) ♦ Pencils	♦ Large grid paper ♦ Dry erase boards and markers ♦ Pawn	♦ Graph paper (cm) ♦ Pencils
Suggested Activity Directions	Place 2 stickers (on various coordinate origins, not spaces) on each piece of graph paper. Demonstrate how to draw a path from one sticker to the other with your finger. Travel on lines (not in spaces) and move across before moving up. Ask students to do the same on their pieces of graph paper and use markers to draw the paths as they explain the directions (e.g., over 3 and up 4). Help students label the coordinates on the graph paper. Repeat.	Draw the *x* and *y* axes on the graph paper for each student. Prompt discussion about the names of each line. Help students write the coordinate labels using numbers for the *x* axis and letters for the y axis. Call out: "Crazy coordinates 4, D." Students search for the coordinates and place their pencil tips on the correct origins. Allow students to take turns calling out different "crazy coordinates." *Coordinate Grid* 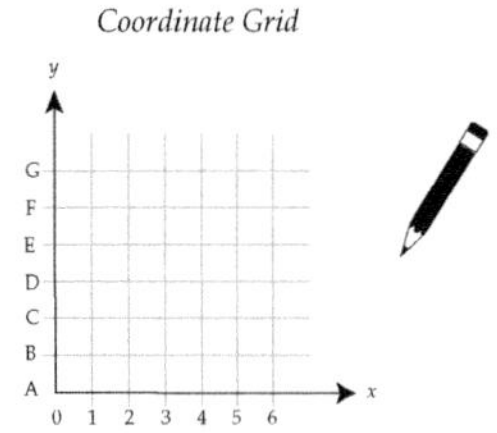 	Draw and label the *x* and *y* coordinates on the large grid paper. Demonstrate how to write coordinates (*x* first followed by *y*, separated by a comma). Distribute dry erase boards and markers. Ask students to close their eyes while you place the pawn on specific origins. Say: "Fast coordinates." Ask students to open their eyes, look at the pawn, and write the coordinates on their dry erase boards. The first student to write the correct coordinates places the pawn next. Play several rounds.	Demonstrate how to draw and label the upper quadrants of the Cartesian coordinate graph (with zero at the bottom center). Ask students to draw the same on their pieces of graph paper. Discuss the value of negatives and highlight the symmetrical positions and distances. Provide time for students to find and record various coordinates. Call out a few coordinates for students to locate. Allow students to call out coordinates for others to find. 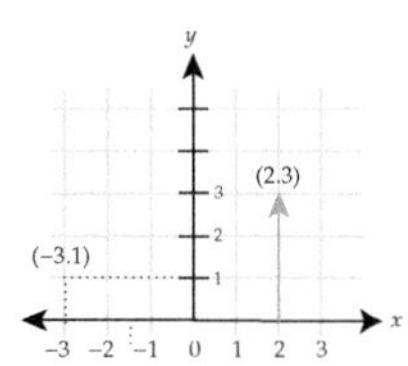
Questions to Assess	♦ How can you make a path? ♦ How should you label the coordinates?	♦ What are the coordinates? ♦ How can you find the location?	♦ What are the coordinates? ♦ How can you describe the origin?	♦ Where is the origin? ♦ What are negative coordinates?

Geometry: Reflective Symmetry

Traditional Formative Assessment

Circle the shapes that are symmetrical.

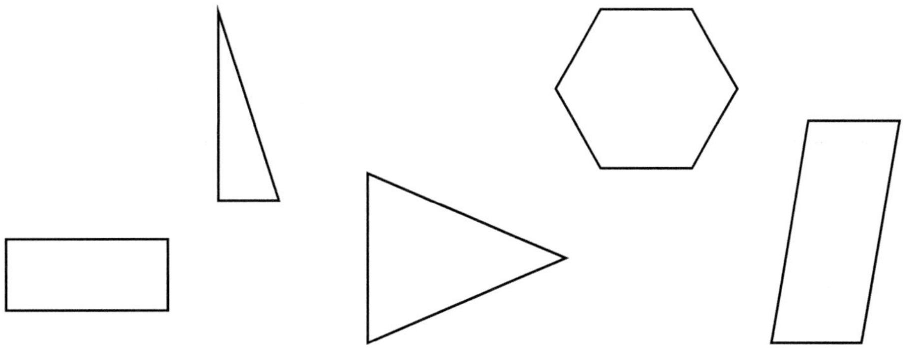

Limitations: Students may just guess and obtain the correct answers.

Enhanced Formative Assessment

Draw the lines of symmetry on each shape that is symmetrical.

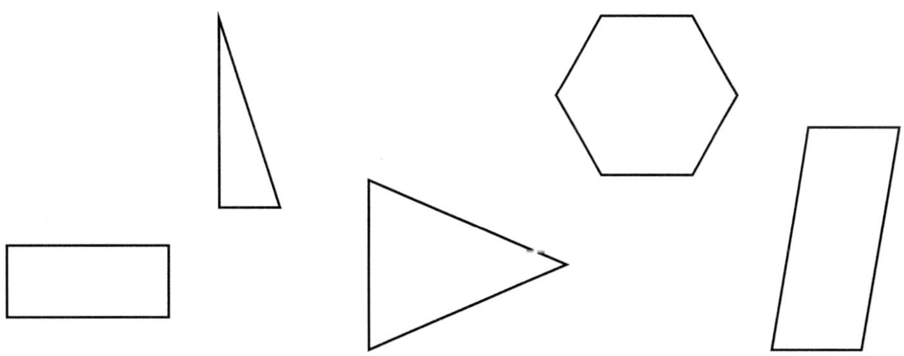

Geometry: Reflective Symmetry

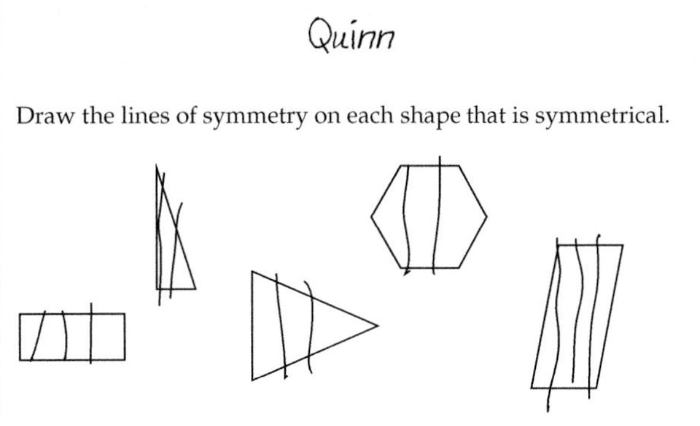	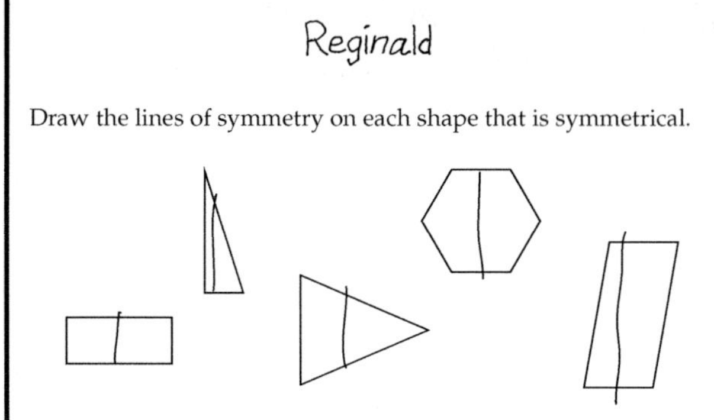
Quinn draws several lines on each shape. The lines do not show symmetry. Instructional Focus: **Rebuild**	Reginald accurately draws a line of symmetry on the rectangle and the hexagon. He does not draw the correct line of symmetry on the isosceles triangle. The right triangle and the parallelogram are not symmetrical. Instructional Focus: **Core 1**
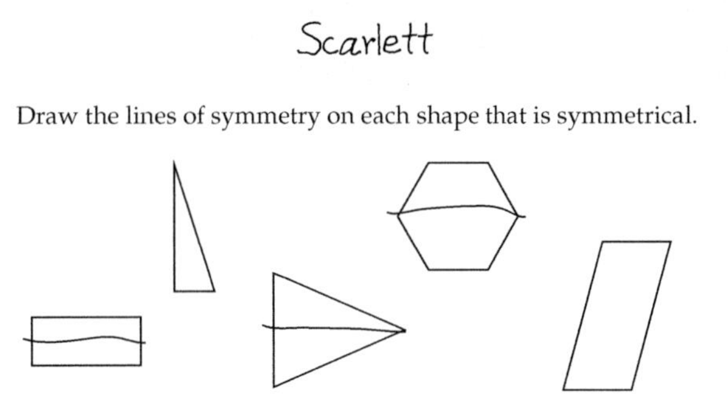	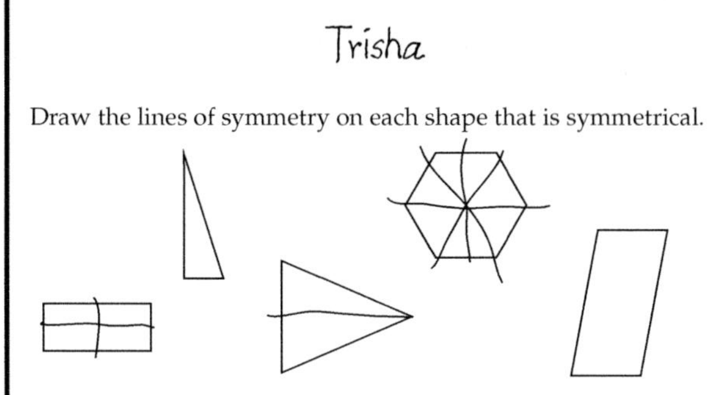
Scarlett accurately draws 1 line of symmetry on the symmetrical shapes. She does not include more than 1 line of symmetry. Instructional Focus: **Core 2**	Trisha draws most of the lines of symmetry on the symmetrical shapes. Instructional Focus: **Challenge**

Geometry: Reflective Symmetry

	Rebuild Focus	Core 1 Focus	Core 2 Focus	Challenge Focus
	Butterflies	**Symmetrical or Not**	**Geo-symmetry**	**Irregular Polygons**
Goal	Explore simple bilateral symmetry	Distinguish between symmetrical and asymmetrical designs	Find more than one line of symmetry	Find lines of symmetry in irregular polygons
Materials	♦ Paper ♦ Paint (different colors) ♦ Spoons	♦ Pattern blocks ♦ Yarn ♦ Tape	♦ Geoboards ♦ Geobands ♦ Long pipe cleaners	♦ Graph paper ♦ Highlighters ♦ Scissors
Suggested Activity Directions	Each student folds a piece of paper in half. Place 2 or 3 small dollops of paint on one side of the paper (toward the center). Ask students to predict what will happen when they close the paper and press the sides together. Introduce the terms symmetry and symmetrical. Students open their papers and investigate the symmetrical butterflies. 	Tape both ends of a piece of yarn on the table in front of each student. Ask students to use pattern blocks to construct a design on one side of the yarn. Students move to another person's design and construct the other side of the design, making it symmetrical. Try another design, but make it asymmetrical. Continue creating and naming symmetrical and asymmetrical designs. 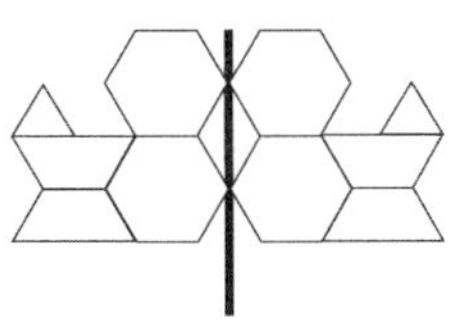	Ask students to make a square with a geoband on their geoboards. Demonstrate how to place a pipe cleaner across the square to depict a line of symmetry. Students do the same. Ask: "Are there any other lines of symmetry?" Help students place additional pipe cleaners to show the 4 lines of symmetry. Repeat with other regular shapes. 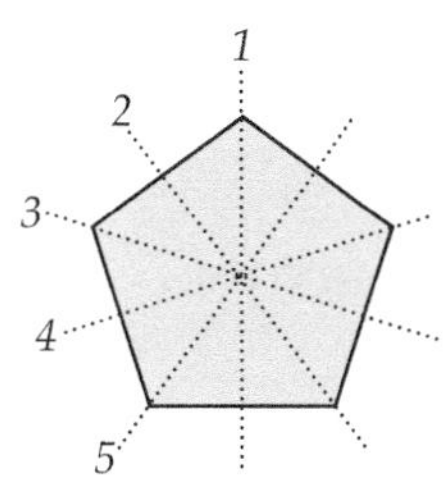 	Explain to students that irregular polygons have sides that are not the same size. Ask students to draw an irregular hexagon on graph paper. Help students use highlighters to draw the lines of symmetry if there are any. Cut out shapes and use folding to verify lines of symmetry. Try irregular pentagons, octagons, heptagons, nonagons, and decagons. 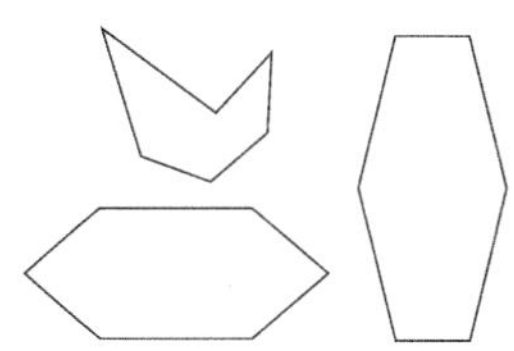
Questions to Assess	♦ How can you describe symmetry? ♦ How are the sides the same?	♦ Is it symmetrical? ♦ How do you know that it is asymmetrical?	♦ How many lines of symmetry can you find in a regular pentagon or a regular octagon?	♦ Is the irregular shape symmetrical? ♦ How do you know?

Geometry: Flips, Slides, and Turns

Traditional Formative Assessment

Draw lines to match the words to the correct definitions.

Flip	Translation
Slide	Rotation
Turn	Reflection

Limitations: Knowledge of the definition does not mean the student understands the concept. Likewise, a student may understand the concept and not know the definition.

Enhanced Formative Assessment

Draw a flip of the arrow.

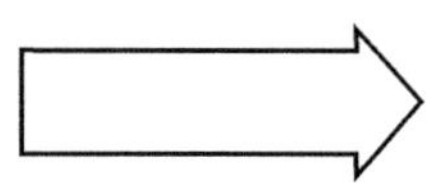

Draw a slide of the arrow.

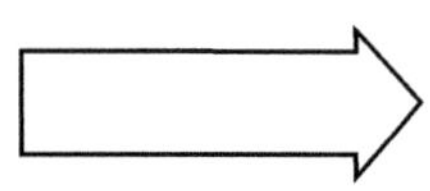

Draw a turn of the arrow
(turn defined as 90 degrees).

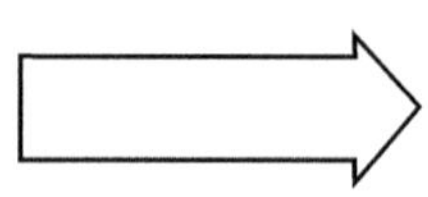

Geometry: Flips, Slides, and Turns

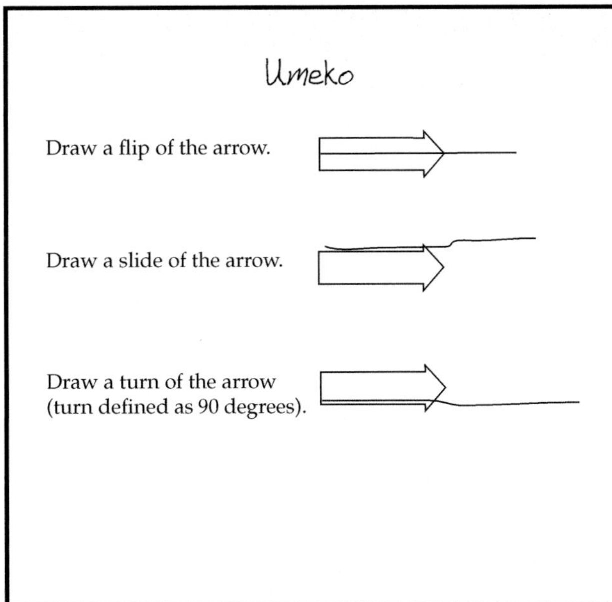

Umeko did not accurately draw any of the transformations.

Instructional Focus: **Rebuild**

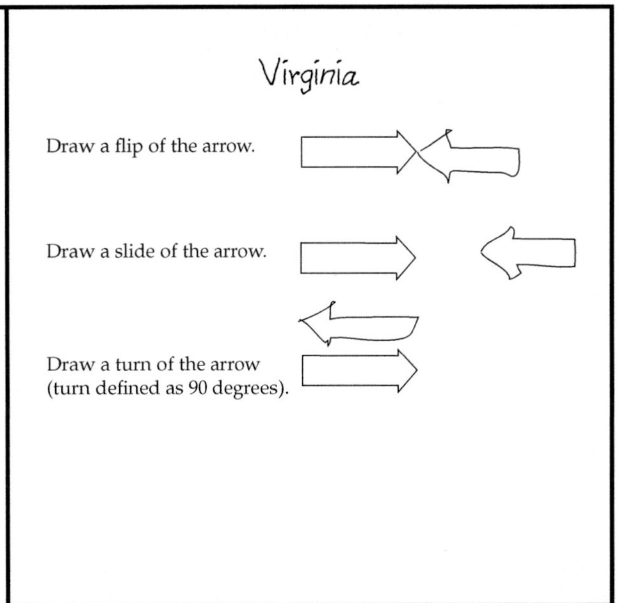

Virginia accurately draws the flipped arrow.

Instructional Focus: **Core 1**

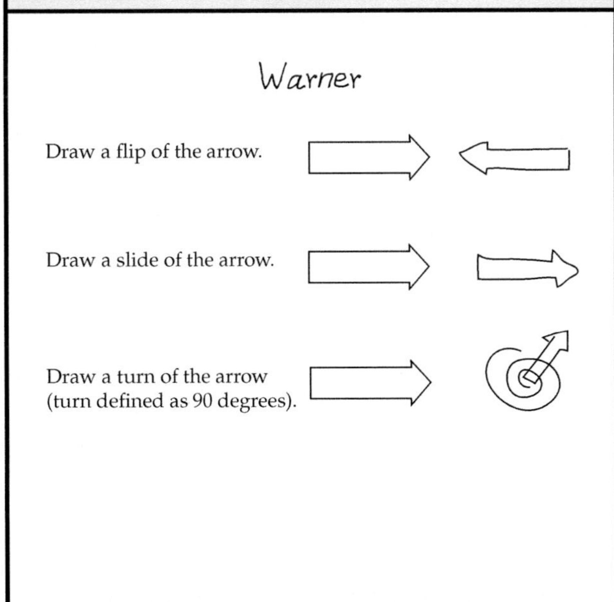

Warner draws the flip and slide correctly. He depicts a spinning arrow to represent the turn.

Instructional Focus: **Core 2**

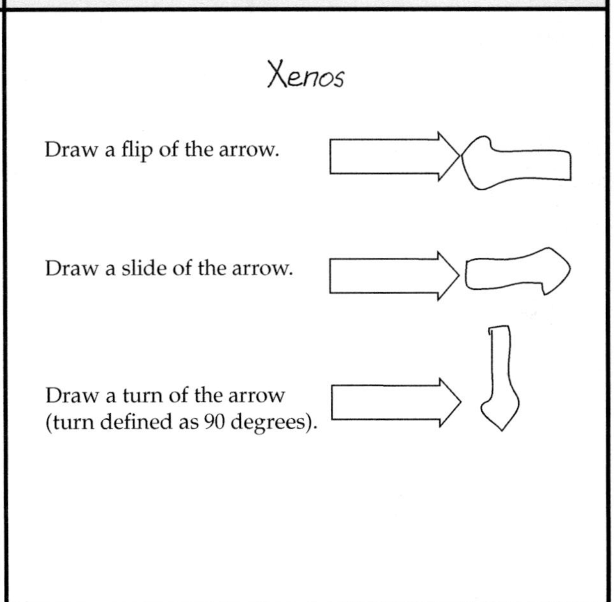

Xenos correctly draws each of the transformations.

Instructional Focus: **Challenge**

Geometry: Flips, Slides, and Turns

	Rebuild Focus	Core 1 Focus	Core 2 Focus	Challenge Focus
	Flip Out!	**Slide Over**	**Your Turn**	**Glide Reflection**
Goal	Understand reflections (flips)	Understand translations (slides)	Understand rotations (turns)	Understand a more complex transformation (glide)
Materials	♦ Grid paper (inch) ♦ Paper squares (1 inch with triangle on front and back) 	♦ Grid paper (inch) ♦ Paper squares (1 inch with smiley face on front and back) 	♦ Grid paper (inch) ♦ Paper squares (1 inch with crescent on front and back) 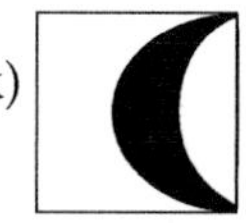 ♦ Pencils	♦ Cutout footprints ♦ Bulletin board paper
Suggested Activity Directions	Distribute grid paper and squares. Ask students to place 1 square inside one of the squares on the grid paper. Explain that reflections are flips. Model how to flip the square to the right. Students show the flip. Demonstrate how to flip to the left and have students show the flip. Repeat with flips up and down. Call out various flip directions (e.g., "flip to the left 3 times"). Encourage students to give flip directions to one another.	Distribute grid paper and squares. Ask students to place one square inside one of the squares on the grid paper. Explain that translations are slides. Model how to slide the square to the right, left, up, and down. Students try the translations. Ask: "How are slides different from flips?" Students show flips and slides. Encourage students to draw and label several slides and flips on the grid paper. 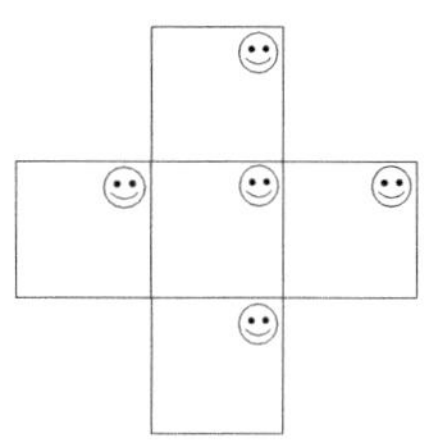	Explain that turns need to be defined. Share that you will work with 90° turns during this lesson. Distribute grid paper and squares. Students place 1 square inside one of the squares on the grid paper. Model how to turn the square 90° to the right using a pencil point to hold the bottom right corner as the square is turned. Students show the turn. Model a 90° turn to the left and have students show the turn. Repeat with 90° turns up and down. Students demonstrate flips, slides, and turns while others guess the transformation type and direction.	Explain to students that a glide reflection (also called a walk) is a 2-step transformation involving a flip and a slide. Ask students to try this type of transformation with one of the footprints. Students trace the footprints after each glide reflection. Ask students to decide if the order (flip then slide) changes how the glide reflection looks. 
Questions to Assess	♦ What does it look like after one flip to the right? ♦ How many flips?	♦ How do you know that is a slide? ♦ Which way did you slide it?	♦ Which way did you turn it? ♦ What are transformations?	♦ Why is the glide reflection also called a walk?

5

mass capacity area

Measurement

time weight length

CCSS Measurement: Length

Traditional Formative Assessment

How long is the pencil?

- O 7/8 inches
- O 7 inches
- O 8 inches
- O About 8 inches
- O 8 ½ inches
- O About 9 inches

Limitations: Showing only part of the ruler and pencil in the graphic can be confusing to students (even to the students who know how to measure in inches).

Enhanced Formative Assessment

Note: This assessment can be modeled with objects or completed on paper.

How can you measure this pencil?

Agustin

How can you measure this pencil?

Agustin does not include the use of any measuring tools. His statement does not include the comparison (long compared to what?).

Instructional Focus: **Rebuild**

Bela

How can you measure this pencil?

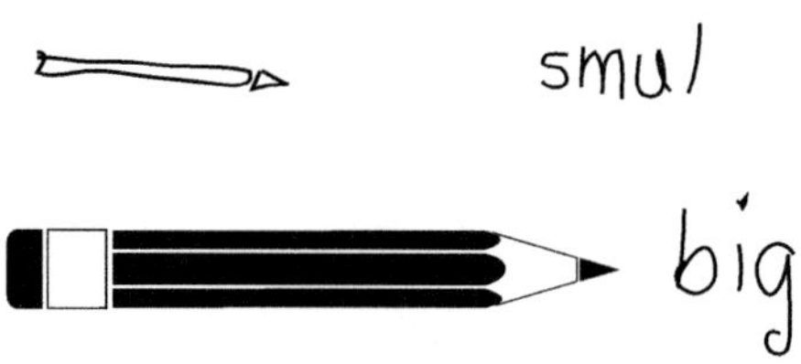

Bela draws a smaller pencil to compare to the original pencil. She clearly labels "small" and "big" to show the comparative size of the pencils.

Instructional Focus: **Core 1**

Camira

How can you measure this pencil?

Camira uses nonstandard units to measure the pencil. She lines the cubes up accurately and records the length in units.

Instructional Focus: **Core 2**

Darren

How can you measure this pencil?

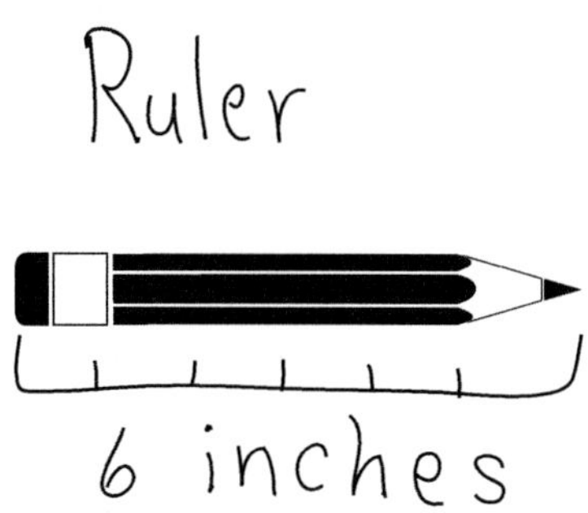

Darren uses a ruler to measure the pencil and then draws a picture of the ruler to show the length. He measures in inches.

Instructional Focus: **Challenge**

CCSS Measurement: Length

	Rebuild Focus	Core 1 Focus	Core 2 Focus	Challenge Focus
	Order the Tubes	**Line up the Cubes**	**Measuring Straws**	**Measuring Books**
Goal	Compare and order lengths	Measure lengths using nonstandard units	Measure lengths using standard units (inches and centimeters)	Measure lengths using standard units (half and quarter inch)
Materials	♦ Paper towel and toilet paper tubes (cut to various lengths) ♦ Paper and crayons	♦ Cubes ♦ Strips of colored paper (various lengths) ♦ Paper and crayons	♦ Simple rulers (with inch side and cm side) ♦ Straws cut in various lengths ♦ Scissors ♦ Numeral cube (die)	♦ Rulers (with half and quarter inch) ♦ Sticky notes ♦ Pencils ♦ Books (various sizes)
Suggested Activity Directions	Place 2 tubes on the table in a vertical position. Prompt discourse about the comparative heights. Repeat with other tubes. Encourage vocabulary such as taller and longer. Introduce 3 tubes and ask students to place these in order from shortest to tallest. Encourage students to explain and justify the placements. Repeat the activity with 4 or 5 tubes. Invite students to draw and describe a set of tubes in order. 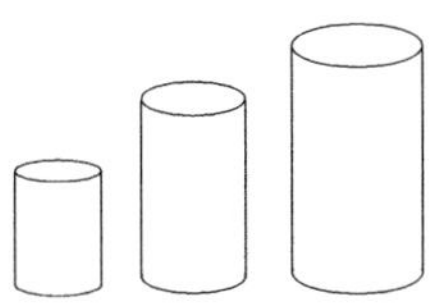	Show students 2 strips of paper (same size). Line cubes up beside one strip to accurately measure the length. Beside the other strip of paper place cubes unevenly spaced and not lined up at the ends. Ask: "Which one shows the correct way to measure the length?" Guide students as they explain why the cubes need to line up and be evenly spaced. Allow students to use cubes to measure various strips of paper. If appropriate, ask students to record the measurements. 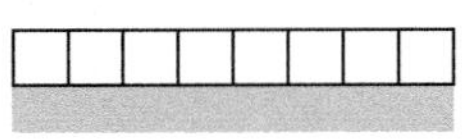	Distribute rulers and help students examine the tick marks, numbers, and labels. Highlight the centimeter side and the inch side. Distribute straws and have students use the rulers to measure the lengths to the nearest inch or centimeter. Help students line up and read the ruler accurately. Ask students to use unit names when they report lengths. Roll the numeral cube and announce the numeral and a unit (in or cm). Ask students to cut straws that fit this measurement. Compare straws for equal lengths. Repeat.	Distribute rulers and help students examine the inches and fractions (half inch and quarter inch). Help students notice that 2 half-inches equal 1 inch and 4 quarter-inches equal 1 inch. Demonstrate how to measure the length and width of a book. Record these measurements on sticky notes and place them on the book. Students each measure a book in the same manner. Ask students to trade books to verify the measurements. Introduce measuring the depth of the book if appropriate.
Questions to Assess	♦ Which tube is shorter? ♦ How much taller is this tube compared to that tube?	♦ How many cubes long is the green paper? ♦ Which paper is 5 cubes in length?	♦ Which is longer, 3 inches or 4 centimeters?	♦ What is the height of the book? ♦ How many half-inches are there in 5 inches?

Traditional Formative Assessment

How much does the box weigh?

Limitations: The scale does not include units. Additionally, students may not understand weight, but write 8 because the hand of the dial is directed to 8.

Enhanced Formative Assessment

Note: Show students an empty pan balance if they are unfamiliar with this measurement tool.

Draw a pan balance to show that 1 eraser weighs the same as 4 cubes.

CCSS Measurement: Weight

Edwina

Draw a pan balance to show that 1 eraser weighs the same as 4 cubes.

Edwina draws 4 erasers, each with 1 cube on top.

Instructional Focus: **Rebuild**

Fulton

Draw a pan balance to show that 1 eraser weighs the same as 4 cubes.

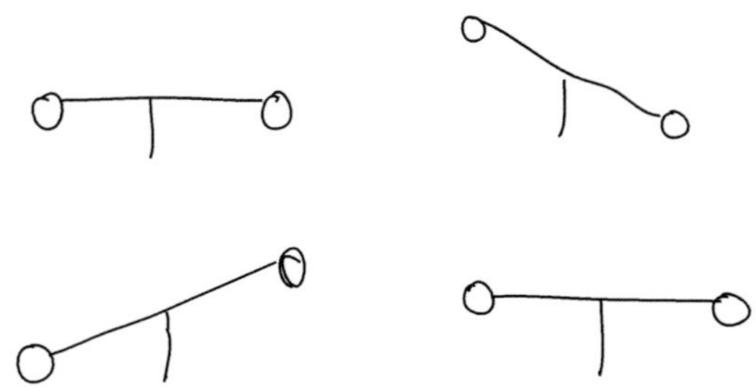

Fulton draws 4 pan balances all showing different comparative weights.

Instructional Focus: **Core 1**

Garrison

Draw a pan balance to show that 1 eraser weighs the same as 4 cubes.

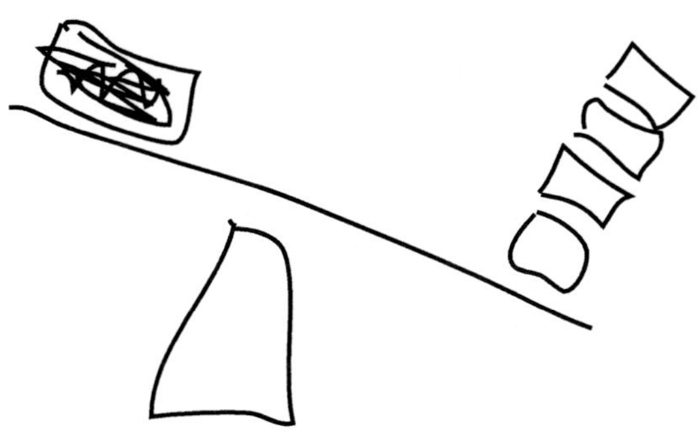

Garrison draws an eraser in one pan and 4 cubes in the other pan. He depicts that the cubes weigh more than the eraser.

Instructional Focus: **Core 2**

Hope

Draw a pan balance to show that 1 eraser weighs the same as 4 cubes.

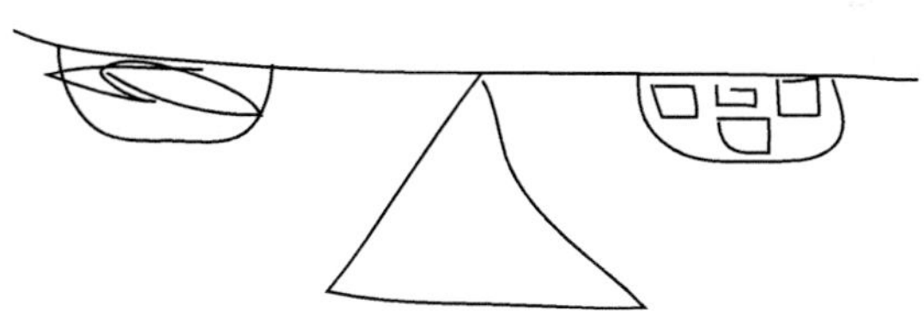

Hope depicts an eraser in one pan and 4 cubes in the other pan. She shows that the weight is equal because it is balanced.

Instructional Focus: **Challenge**

CCSS Measurement: Weight

	Rebuild Focus	Core 1 Focus	Core 2 Focus	Challenge Focus
	Balancing Bears	**Order the Boxes**	**What's the Weight?**	**How Many Grams?**
Goal	Compare the weight of items using a pan balance	Order items by weight using the pan balance	Measure weight with nonstandard units using a balance	Measure weight with standard units using a platform scale
Materials	♦ Pan balances ♦ Counting bears (same size and weight)	♦ Pan balances ♦ Small boxes with various amounts of pennies in each	♦ Bucket balances ♦ Objects to weigh ♦ Cubes ♦ Paperclips	♦ Platform scales ♦ Gram weights
Suggested Activity Directions	Introduce the pan balance and prompt discussion about how it functions. Place a bear in one pan. Ask students to describe how the pan balance looks (the side with the bear goes down). Move the bear to the other pan and encourage discussion. Try placing a bear in each pan and describing the comparative (equal) weights. Place 2 bears in one pan and 1 bear in another pan. Discuss how the pan balance indicts which side is heavier or lighter. Give each student a pan balance and some bears. Encourage students to try comparing the weights of different bears.	Show 3 boxes (same size, but different weights) and a pan balance. Ask students how to use the balance to find out which box has the greatest weight. Help students compare the weights and identify the heaviest box. Compare the other 2 boxes on the balance. Encourage students to use the information to place the 3 boxes in order from heaviest to lightest. Give each student 3 different boxes to compare and order by weight. Try more than 3 boxes if appropriate. 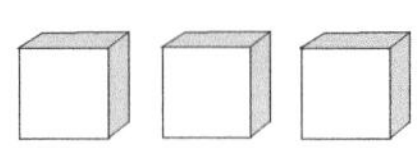	Demonstrate how to weigh an object with nonstandard units (cubes). Encourage discussion about how the balance shows equality. Ask students to record the weight. For example, 1 pencil weighs 3 cubes. Give each student a bucket balance, some cubes, and an object. Students find and record the weight of the object. Provide students the opportunity to verify the weights found by others. Try a different nonstandard unit (such as paperclips). Compare the weights. 	Encourage students to explore with the gram weights on the platform scales by placing a given number of weights and reading the scale. Announce various weights and have students display these weights on the platform scales. Encourage students to notice various ways to show the weights (e.g., 25 grams can be shown with five 5-gram weights or two 10-gram weights and one 5-gram weight). Record weights if appropriate. 
Questions to Assess	♦ Which side is heavier? How do you know? ♦ How can you show balance?	♦ Which box is lightest? ♦ How can you order the boxes by weight?	♦ How many cubes does the crayon weigh? ♦ What if you add one more cube?	♦ How can you show 15 grams? ♦ How many grams are you showing?

Traditional Formative Assessment

This jar will hold ____ ping pong balls.

- O 2
- O 5
- O 17
- O 25

Limitations: The assessment fails to include the point of reference (size of the ping pong ball). Furthermore, the jar's dimensions are unknown.

Enhanced Formative Assessment

Show students 5 ping pong balls inside of a jar that will hold about 20.

Say: "Estimate how many ping pong balls the jar will hold."

Measurement: Capacity

Isidore	Jana
"Five."	"One hundred."
Isidore names the number of ping pong balls currently in the jar. He does not estimate how many ping pong balls the jar will hold. Instructional Focus: **Rebuild**	Jana's estimation is not reasonable. She does not use the referent (5 balls). Instructional Focus: **Core 1**
Kelvin	**Lexie**
"There are five in there already. I think about fifty will fit."	"Well, five takes up about half of the bottom half. So, five more would be ten. Another group of ten makes twenty. I estimate twenty."
Kelvin uses the referent, but his estimation is not within a good range. Instructional Focus: **Core 2**	Lexie uses an efficient strategy to estimate the capacity with nonstandard units. Instructional Focus: **Challenge**

Measurement: Capacity

	Rebuild Focus	Core 1 Focus	Core 2 Focus	Challenge Focus
	Cubes in a Cup	**Cups in Order**	**Spoon the Popcorn**	**A Gallon of Water**
Goal	Compare and estimate capacities	Order and estimate capacities using referents	Use range-based estimation of capacities (nonstandard units)	Convert and estimate cups, quarts, and gallons
Materials	♦ Clear cups (two different sizes) ♦ Cubes	♦ Clear cups (various sizes) ♦ Bear counters ♦ Note cards	♦ Three jars (different sizes) ♦ Unpopped popcorn ♦ Spoon	♦ Gallon jug of water ♦ 4 quart containers ♦ 8 pint containers ♦ 16 cups ♦ Funnel
Suggested Activity Directions	Place cubes in a cup one at a time. Students count the cubes as they are placed in the cup. Discuss how full the cup is at several points during the process (e.g., not full, half full, almost full). Remove cubes one at a time, counting down and describing the capacities (include close to empty and empty). Show a different-sized cup and help students use the information that they gathered with the first cup to estimate how many cubes will fit into the second cup. Students find the actual capacity by placing cubes in the cup. Discuss which estimates were in a close range to the actual capacity.	Display the various cup sizes. Ask students to estimate which cup will hold the most. Discuss which cup will hold the least. Prompt discussion about where the other cups fit in order. Fill the smallest cup with bear counters. Encourage students to count the bears as they are placed in the cup. Write the capacity on a note card and place it beside the cup. Pour the bears from the smaller cup into the next cup. Ask: "How many bears are needed to fill the cup?" Count on while adding bears to fill the cup. Record the capacity. Continue with the rest of the cups. Adjust order if necessary.	Fill the smallest jar with spoonfuls of popcorn. Encourage students to count the number of spoonfuls it takes to fill the jar. Use the known quantity as a referent to estimate the number of spoonfuls it will take to fill the other jars. Prompt a discussion about range-based estimation by asking students to name a good range for the estimation and to explain why the range is good. Discuss estimates that are not in a good range and why. Remind students that estimations in a good range are correct answers. 	Display gallon, quarts, pints, and cups. Discuss sizes and unit names. Pour water from the gallon into the 4 quarts. Discuss equivalency (1 gallon = 4 quarts). Ask students to estimate how many pints equal 1 gallon. Pour water from the quarts into the pints. Estimate how many cups equal 1 gallon. Pour water from the pints into the cups. Pour all 16 cups of water back into the gallon. Ask students to think about how many cups, pints, or quarts equal 2 gallons. Provide time for students to search the classroom to find a container that holds about 2 gallons. Test the container.
Questions to Assess	♦ How many cubes will fill it? ♦ How many more cubes do you need?	♦ How many more bears do you need to fill the cup? ♦ Is the cup full?	♦ How many spoonfuls will fill the jar? ♦ What if the jar was half of the size?	♦ How many quarts in a gallon? ♦ How many pints in two gallons?

Measurement: Area

Traditional Formative Assessment

Is the area of this rectangle equal to 5?

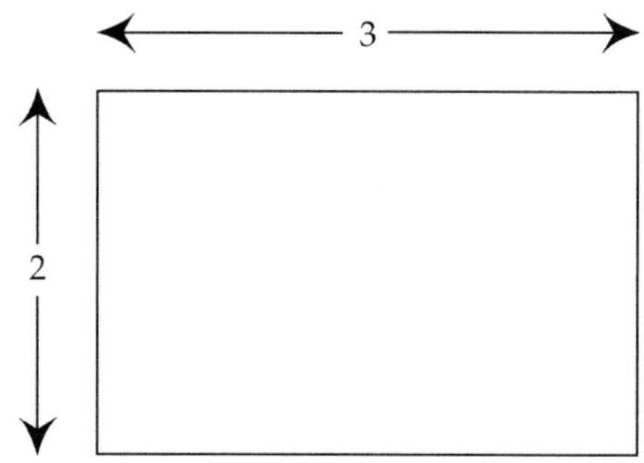

Limitations: The units of measure should be included. Additionally, students may not be familiar with how to use multiplication to find the area.

Enhanced Formative Assessment

Let's work with area.

How many squares will fit in the rectangle?

Explain your answer with words, numbers, or pictures.

Measurement: Area

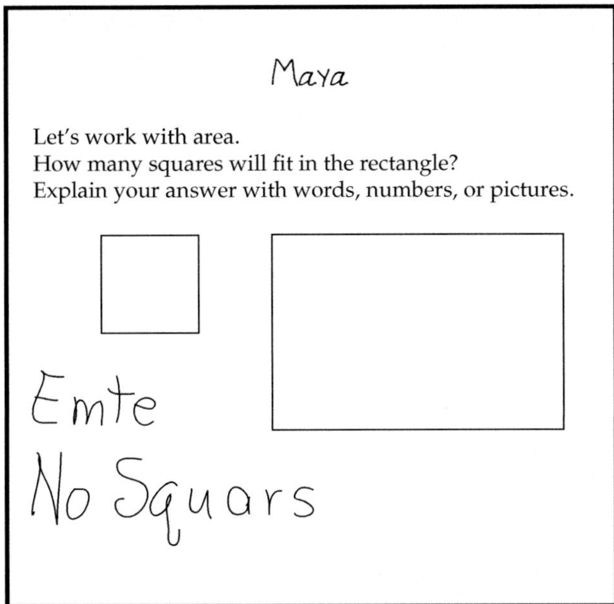

Maya reports that it is empty because there are no squares inside of the rectangle.

Instructional Focus: **Rebuild**

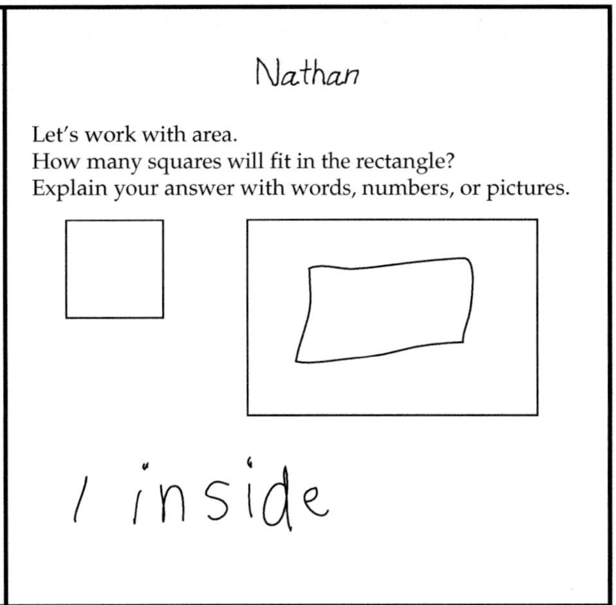

Nathan draws 1 square inside of the rectangle.

Instructional Focus: **Core 1**

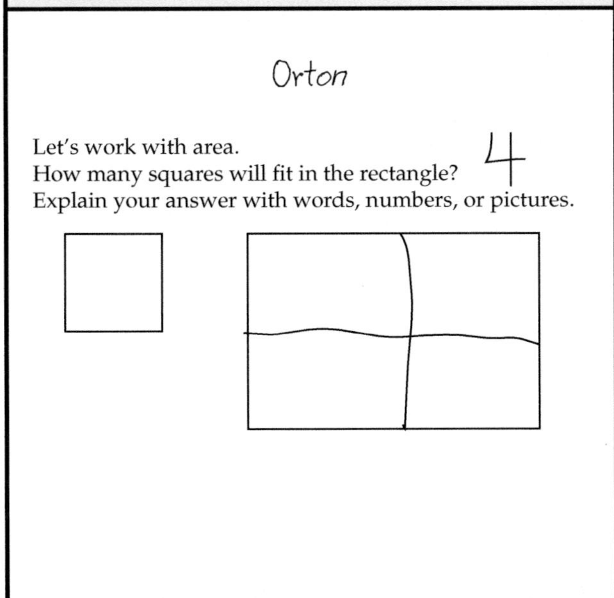

Even though Orton's answer is incorrect, his response indicates that he has some understanding of area.

Instructional Focus: **Core 2**

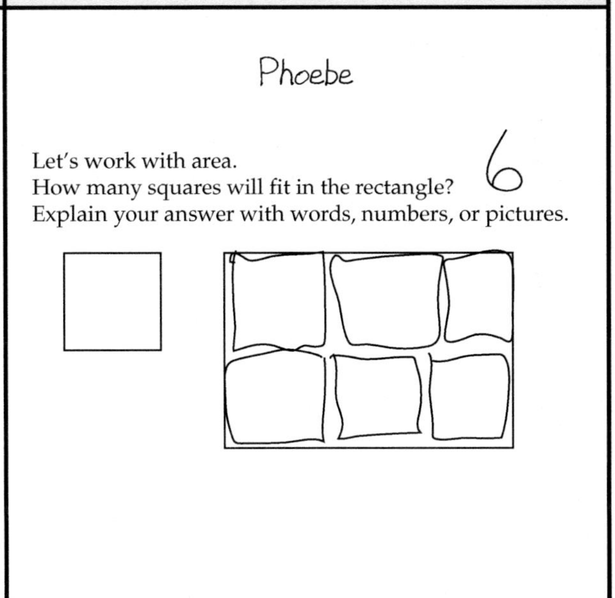

Phoebe accurately identifies the area and explains her answer with pictures of squares correctly placed inside of the rectangle.

Instructional Focus: **Challenge**

Measurement: Area

	Rebuild Focus	Core 1 Focus	Core 2 Focus	Challenge Focus
	Let's Make Rugs!	**Tablecloths**	**Pentominoes**	**Area of a Triangle**
Goal	Compare and estimate area of rectangles	Order and estimate area of rectangles	Create different figures with the same area	Measure area of triangles
Materials	♦ Paper ♦ Scissors ♦ Crayons	♦ Paper rectangles (different sizes, inch dimensions) ♦ Tiles (1 inch)	♦ Paper squares (1 inch) ♦ Tape	♦ Paper triangles (different sizes, inch dimensions) ♦ Graph paper (inch) ♦ Markers
Suggested Activity Directions	Present the idea of making miniature rectangular rugs out of paper. Explain that each student will design a rug on paper and cut it out. After the rugs are completed they are to compare rugs. Ask students to find a rug that covers a smaller area and a rug that covers a larger area than their own. Encourage students to state area comparison. Ask: "Do any of the rugs cover an area that is similar in size?" 	Explain to students that each paper rectangle represents a miniature tablecloth. The task is to compare the areas of all the tablecloths and place them in order. Help students cover the tablecloths with tiles to find the areas. Encourage students to estimate how many tiles will cover the tablecloth before finding the actual area. If needed, students can draw the tiles or write the area on the tablecloths. 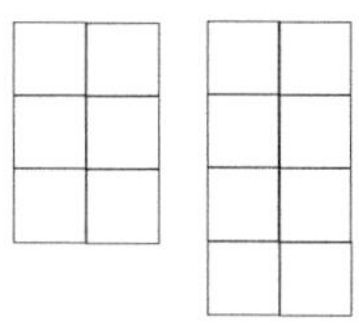	Give each student 5 paper squares. Explain that each student will tape together the 5 squares (edge to edge) to construct different figures that all have the same area. When students finish the task, ask them to try another figure. Flip and turn figures to see how they are similar and different. Ask: "How many different ways can the squares be arranged?" (Hint: There are 12 different pentominoes.)* 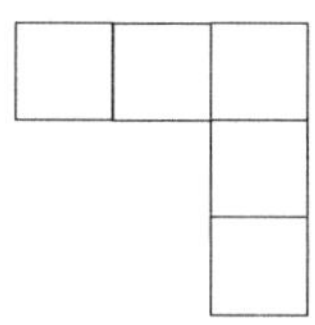	Begin with right triangles. Ask each student to place a triangle on graph paper, matching the triangle's base on a line. Demonstrate how to draw a rectangle using the triangle's height and width dimensions. Discuss the equivalency of the two triangular halves of the rectangle. Help students use the graph paper squares to find the area of the rectangle and the triangle (half of the rectangle). Ask students to find the area of other triangles, including equilateral triangles.
Questions to Assess	♦ Is the area of your rug larger than the other rug? ♦ Which rug covers the smallest area?	♦ What is the area of the tablecloth? ♦ How many tiles do you think will fit?	♦ What is the area? ♦ What if we try to make a 6 × 10 rectangle using all 12 pentominoes?	♦ How can you find the area of a triangle? ♦ What if the triangle was 2 inches taller?

* A similar version of this activity appears in Taylor-Cox, *Family Math Night: Math Standards in Action* (Larchmont, NY: Eye On Education, 2005).

Traditional Formative Assessment

What time is it?

Limitations: The question should be, "What time does the clock show?" Answers to the question, "What time is it?" will vary based on what time of day students answer the question. Furthermore, if the student answers correctly, his level of understanding time beyond one-hour intervals is still unknown.

Enhanced Formative Assessment

Draw the hands on the clocks.

Show 7:00. Show 10:30. Show 1:45.

CCSS Measurement: Time

Quintus

Draw the hands on the clocks.

Show 7:00. Show 10:30. Show 1:45.

Quintus circles the numbers on the clocks that correspond to the hours. He does not draw the hands of the clocks.

Instructional Focus: **Rebuild**

Ralph

Draw the hands on the clocks.

Show 7:00. Show 10:30. Show 1:45.

Each of Ralph's clocks shows a time on the hour.

Instructional Focus: **Core 1**

Shantelle

Draw the hands on the clocks.

Show 7:00. Show 10:30. Show 1:45.

Shantelle accurately shows time by the hour and by the half hour.

Instructional Focus: **Core 2**

Taylor

Draw the hands on the clocks.

Show 7:00. Show 10:30. Show 1:45.

Taylor accurately shows all of the times.

Instructional Focus: **Challenge**

CCSS Measurement: Time

	Rebuild Focus	Core 1 Focus	Core 2 Focus	Challenge Focus
	Show the Time	**Choose the Time**	**Every 15 Minutes**	**Time Goes By**
Goal	Understanding time by the hour and elapsed time by the hour	Understanding time by the half hour and elapsed time by the half hour	Understanding time by the quarter hour and elapsed time by the quarter hour	Understanding time by minute intervals and elapsed time by minute intervals
Materials	♦ Geared clocks ♦ Dodecahedron die (or 1–12 spinner)	♦ Geared clocks ♦ Time cards (on the hour and on the half-hour times)	♦ Geared clocks ♦ Dry erase boards and markers	♦ Geared clocks ♦ Time cards (by minute intervals) ♦ Hundred chart ♦ Beans
Suggested Activity Directions	Give each student a clock. Roll the dodecahedron die. Announce the number rolled as an o'clock time. For example, 6:00. Encourage students to show the time on their clocks. Check for accuracy. Continue rolling the die, announcing times to be shown on the clocks. Introduce the idea of 1 hour later. Roll the die, announce the time, and ask the students to show 1 hour later. Encourage students to move the minute hand all the way around the clock to represent 1 hour later. If appropriate, try 2 hours later or 1 hour before the rolled times.	Distribute clocks and place the time cards face down in a pile. Students take turns choosing time cards and showing the times on individual clocks (check for accuracy of the placement of the hour hand). Introduce 1 hour later. When a student chooses a time card, she announces the time and tells what time it will be 1 hour later. After a few turns, try 1 hour before. Then introduce 1 half-hour (30 minutes) later and repeat the activity. Try 1 half-hour (30 minutes) earlier when students are ready. 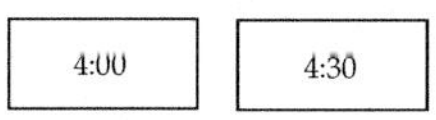 	Give each student a clock and dry erase board. Warm up by discussing, showing, and writing times by the hour and half-hour. Introduce quarter-hour times. Announce, write, and show an o'clock time, quarter after, half past, 45 minutes after, and finally the next hour. Continue going through each hour in this manner. Ask students to say the times as they show the times. When students are ready, introduce the idea of elapsed times (hours, half-hours, and then quarter-hours). 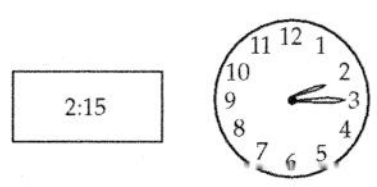 	Students take turns choosing time cards and showing times. Others verify the times shown. After a few rounds, prompt discussion about elapsed time by minute intervals. What time will it be in 10 minutes? What time will it be in 56 minutes? Resume choosing cards and showing times. Add the component of showing the elapsed time when minutes are added. To find out the minutes after (to be added to) each time, drop a bean on the hundred chart. When students are ready, drop the bean to show the minutes before (to be subtracted).
Questions to Assess	♦ Where does the minute hand go? ♦ What time will it be in 3 hours?	♦ How do you know that it is half past the hour? ♦ What time will it be in 30 minutes?	♦ How do you write quarter past 7:00? ♦ Where is the hour hand when it is 3:45?	♦ What time will it be in 83 minutes? ♦ What if it were 3 hours later?

6

Data Analysis & Probability

graphs line plots bar likely events T-charts circle outcomes

Traditional Formative Assessment

Do people like green beans? Yes or No.

I like Green Beans

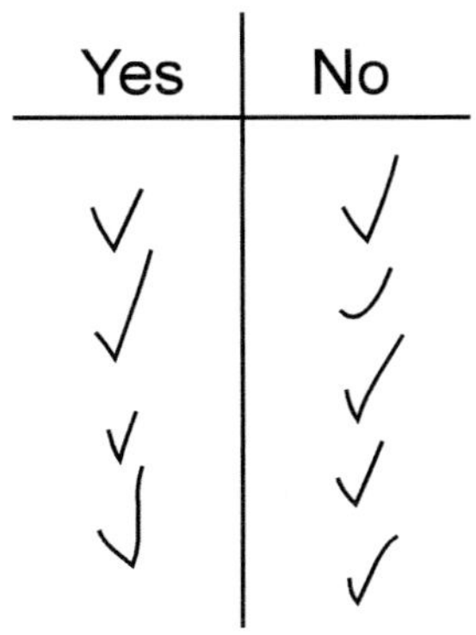

Limitations: The data are organized in a confusing manner. The number of checks in each column is different, yet the amount of space used for each column of data is similar. Additionally, a correct answer could easily be a guess because there are only two possible answers.

Enhanced Formative Assessment

Survey five people and organize the data on this T-chart.
Use dot stickers.

Do You Like Green Beans?

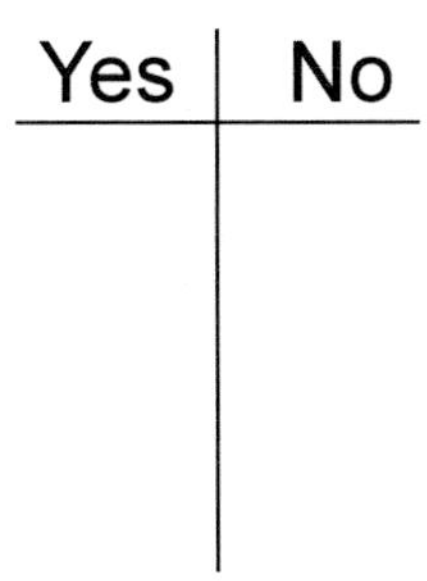

Andre

Survey five people and organize the data on this T-chart.
Use dot stickers.

Do You Like Green Beans?

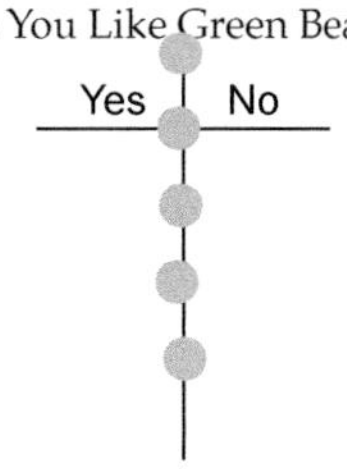

Andre aligns the stickers on the center line. He does not gather or organized the data in a meaningful way.

Instructional Focus: **Rebuild**

Bessie

Survey five people and organize the data on this T-chart.
Use dot stickers.

Do You Like Green Beans?

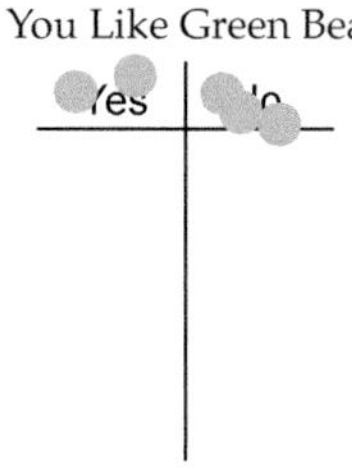

Bessie gathered and organized the data. She is ready to learn how to display the data in a manner that makes it easier to analyze.

Instructional Focus: **Core 1**

Clio

Survey five people and organize the data on this T-chart.
Use dot stickers.

Do You Like Green Beans?

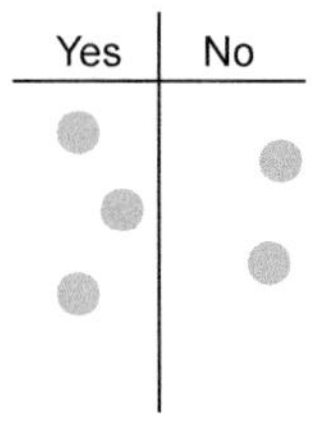

Clio gathers, organizes, and displays the data. She is ready to learn how and why to evenly space the data.

Instructional Focus: **Core 2**

Dennis

Survey five people and organize the data on this T-chart.
Use dot stickers.

Do You Like Green Beans?

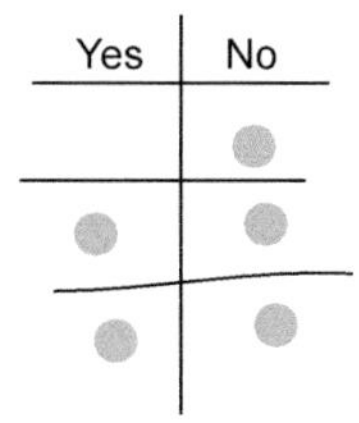

Dennis accurately gathers, organizes, and displays the data. He even draws lines to show how the data are organized.

Instructional Focus: **Challenge**

CCSS Data Analysis and Probability: T-Charts

	Rebuild Focus	Core 1 Focus	Core 2 Focus	Challenge Focus
	Do You Have a Pet?	**Curly or Not?**	**Wearing Blue**	**Pepperoni Pizza**
Goal	Gather and organize simple data in a T-chart	Display simple data in a T-chart	Display greater quantities of data in a T-chart	Work with more complex data in a T-chart
Materials	♦ Tag board T-chart (Yes/No) ♦ Clothespins	♦ Tag board T-chart (Yes/No) ♦ Dot stickers ♦ Paper squares (same color)	♦ Tag board T-chart (Yes/No) ♦ Blue tiles (1 inch)	♦ Tag board T-chart (Yes/No) ♦ Paper circles
Suggested Activity Directions	Ask students if they have any pets. Give each student a clothespin. Ask students to answer the question, "Do you have any pets?," by placing clothespins on the Yes/No T-chart. Encourage students to ask other students the same question. Ask them to place a clothespin to represent each piece of data collected. Encourage students to organize the data in a manner that allows others to clearly see the information. Discuss the results. How many people have pets? How many people do not have pets?	Introduce the survey question: Do you have curly hair? Each student in the group attaches a dot sticker to a paper square. The square is placed on the Yes/No T-chart. Help students align the squares in the correct locations in columns on the T-chart. Prompt discussion about the data. How many people have curly hair? How many people do not have curly hair? If we asked your family to participate in the survey, where would the dot squares be placed?	Introduce the survey question: Are you wearing blue today? Students survey other students in the class. Demonstrate how to use one blue tile for each piece of data collected. Help students line up the tiles (evenly spaced) in columns on the T-chart. Compare the data. Encourage students to make statements about the data (e.g., 8 people are wearing blue today). Ask students to predict what the data might look like tomorrow.	Introduce the survey question: Do you like pepperoni pizza? Students survey other members of the class. Explain that the data will be represented with circles on the Yes/No T-chart. To add a bit of complexity to the task, share that each circle represents 2 people. If the data include an odd number, show students how to display the data by cutting one of the circles in half. Help students create a legend (key) for the chart.
Questions to Assess	♦ What do you notice about the data? ♦ How can we keep the data organized?	♦ Do more people have curly hair? ♦ Where is your piece of data on the T-chart?	♦ How many people are not wearing blue today? ♦ How many more people are wearing blue than not wearing blue?	♦ How many people like pepperoni pizza? ♦ How might the data change if we asked another class to participate in the survey?

Traditional Formative Assessment

Look at the bar graph and mark the correct answer.

Which center choice is the most popular? Writing, Math Blocks, or Books?

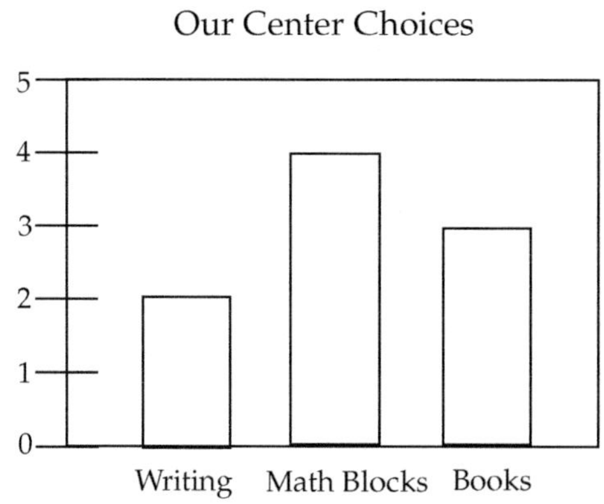

Limitations: Students may simply choose the word with the largest rectangle without having an understanding of bar graphs.

Enhanced Formative Assessment

Fill in the missing parts of this bar graph about our center choices (writing, math blocks, and books.)

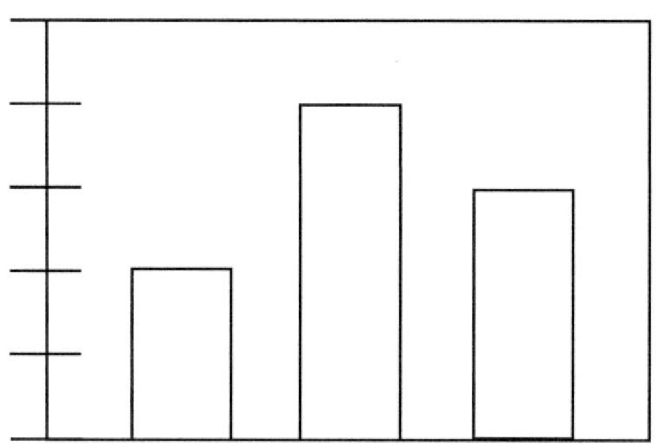

Ellie

Fill in the missing parts of this bar graph about our center choices (writing, math blocks, and books.)

Ellie's "missing parts" include details to create characters, rather than a bar graph.

Instructional Focus: **Rebuild**

Flynn

Fill in the missing parts of this bar graph about our center choices (writing, math blocks, and books.)

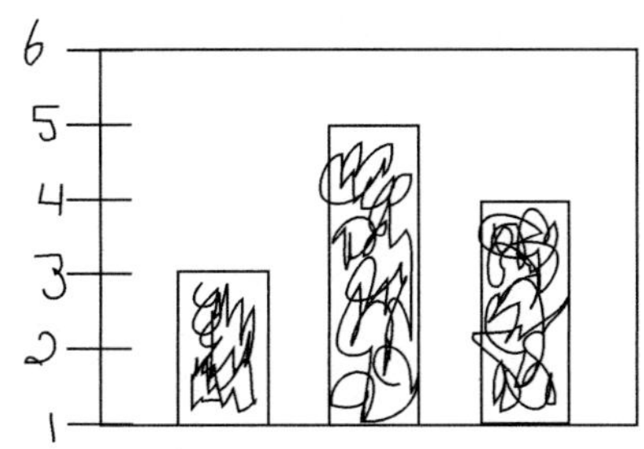

Flynn shades the columns on the bar graph and adds numbers. The scale is incorrect because he does not include zero.

Instructional Focus: **Core 1**

Gordon

Fill in the missing parts of this bar graph about our center choices (writing, math blocks, and books.)

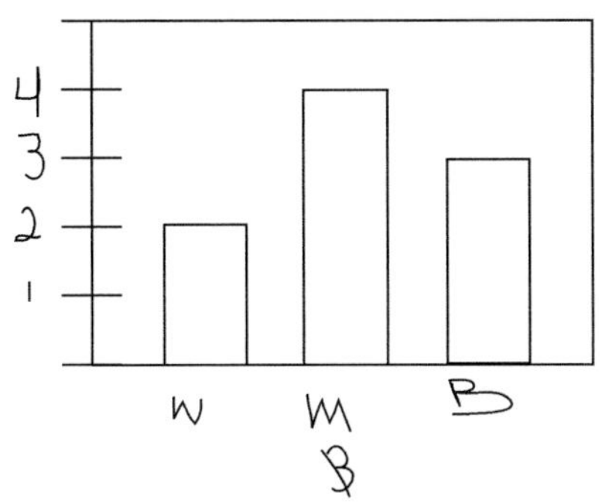

Gordon adds category labels (first letter of each word) and numbers. Even though he does not write zero, his scale is correct.

Instructional Focus: **Core 2**

Heather

Fill in the missing parts of this bar graph about our center choices (writing, math blocks, and books.)

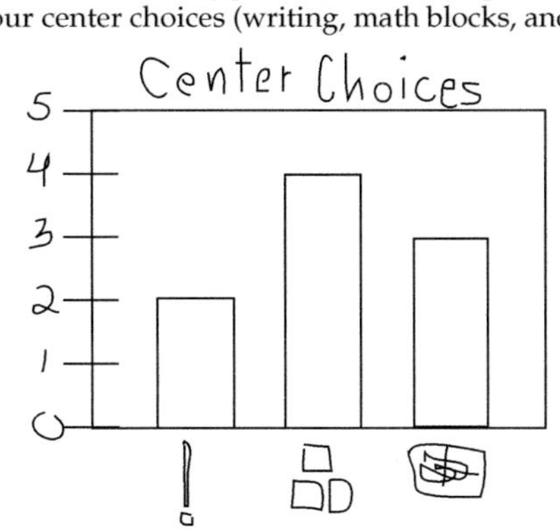

Heather includes category label pictures, correct numbers on the scale, and a title for the graph.

Instructional Focus: **Challenge**

CCSS Data Analysis and Probability: Bar Graphs

	Rebuild Focus	Core 1 Focus	Core 2 Focus	Challenge Focus
	Rock Collection	**Animals at the Zoo**	**Backpacks**	**Cookies for All**
Goal	Discuss data on a real graph (data are represented with actual objects)	Include accurate scales on bar graphs (data are represented with pictures)	Create titles for bar graphs (data are represented with abstract symbols)	Include axes labels on bar graphs and work with pictographs (pictures each represent more than one piece of data)
Materials	♦ Real graph template ♦ Rocks (different textures)	♦ Bar graph template ♦ Pictures of 4 types of zoo animals (on index cards)	♦ Bar graph template ♦ Cubes	♦ Bar graph template ♦ Paper cookies ♦ Glue
Suggested Activity Directions	Prepare a real graph with title, scale, and category labels (without data). Ask: "How can we sort the rocks by texture?" Discuss students' responses. Introduce categories such as coarse-grained, fine-grained, or smooth if necessary. Examine the rocks. Talk about the categories. Prompt discussion about where the data (actual rocks) should be placed on the real graph. Help students accurately line up the data (rocks). Once the data are placed, discuss the data and who may like to know this information.	Prepare a bar graph with title and category labels (without the numeric scale or data). Ask students to sort the picture cards by type of animal. Discuss where to place the animal cards on the bar graph template. Help students accurately align the data (animal picture cards). Focus attention on the missing scale. Ask students to write numbers in the correct location on the bar graph. Discuss how the numbers name the lines (not spaces) and the significance of zero.	Prepare a bar graph with numeric scale (without the title, category labels, or data). Ask students to compare their backpacks and decide how they would like to sort the backpacks. Encourage them to be creative in their categories (beyond color). Write these category names on the bar graph. Explain to students that one cube represents each backpack. Help students place the cubes on the graph. Ask students to think of a tile for the graph.	Prepare a bar graph with title, scale, and category labels (without data). Help students use attendance sheets from at least 3 classes to collect the total number of students present in each class. Share that a pretend baker needs data in order to make 1 cookie for each student. Explain that the baker wants a pictograph with a scale of 1 paper cookie = 2 students. Help students arrange the data, create the legend or key, and add axes labels to the bar graph.
Questions to Assess	♦ How many smooth rocks are in the set? ♦ Do we need another category called "mixed texture?"	♦ Which animal did the children see most often at the zoo? ♦ Who may want to know this information?	♦ How many backpacks are in the first category? ♦ Where is the representation of your backpack?	♦ How can we show the data if a class has an odd number of students?

Traditional Formative Assessment

Which statement is correct about the circle graph?

- O More students ate cereal this morning.
- O More students ate toast this morning.
- O More students drank milk this morning.

This Morning's Breakfast

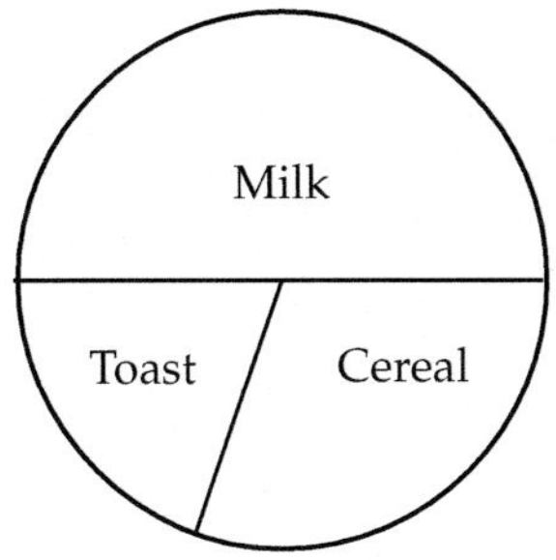

Limitations: In addition to the limitations of multiple choice questions, the circle graph is confusing because some of the categories overlap (one person could be in several categories). More information such as, one or more totals should be provided.

Enhanced Formative Assessment

Make a circle graph with these data.

There are 10 people coming for breakfast.
Five people want cereal.
One person wants pancakes.
Four people want eggs.

Breakfast for Ten

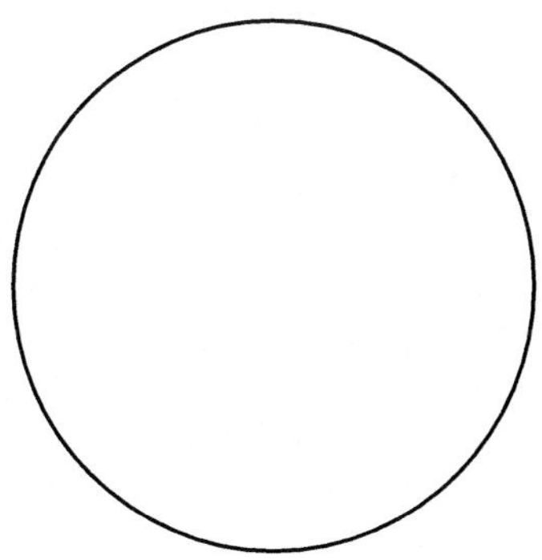

Data Analysis and Probability: Circle Graphs

Iggy

Make a circle graph with these data.

There are ten people coming for breakfast.
Five people want cereal.
One person wants pancakes.
Four people want eggs.

Breakfast for Ten

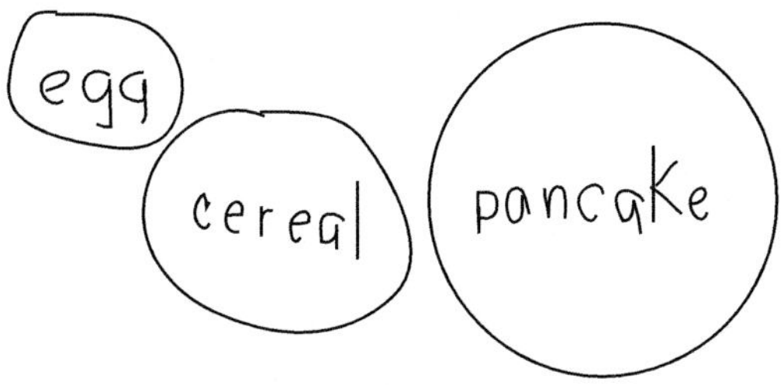

Iggy draws the 3 breakfast options. He does not create a circle graph with the data.

Instructional Focus: **Rebuild**

Jamila

Make a circle graph with these data.

There are ten people coming for breakfast.
Five people want cereal.
One person wants pancakes.
Four people want eggs.

Breakfast for Ten

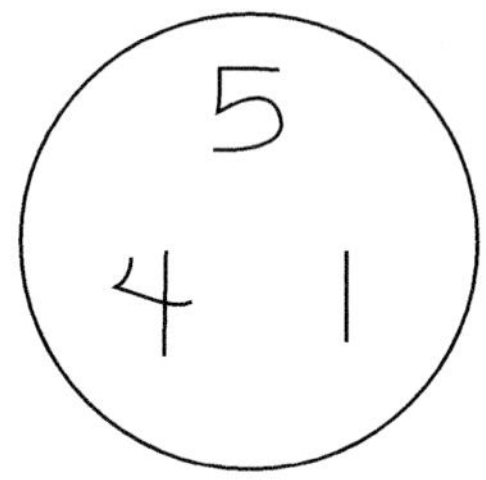

Jamila places the numbers in the circle without showing the sections or category names.

Instructional Focus: **Core 1**

Kane

Make a circle graph with these data.

There are ten people coming for breakfast.
Five people want cereal.
One person wants pancakes.
Four people want eggs.

Breakfast for Ten

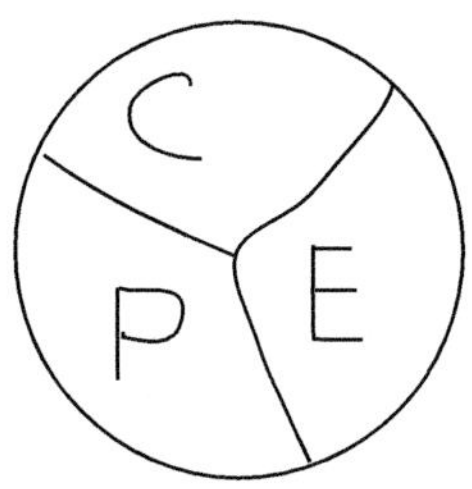

Kane shows a circle graph with 3 equal parts labeled by the first letter of each word.

Instructional Focus: **Core 2**

Lenora

Make a circle graph with these data.

There are ten people coming for breakfast.
Five people want cereal.
One person wants pancakes.
Four people want eggs.

Breakfast for Ten

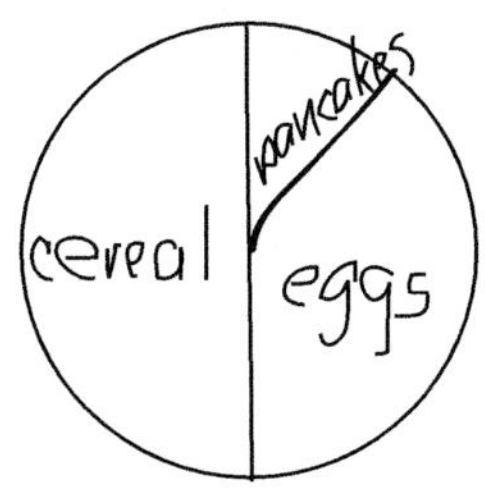

Lenora accurately displays the data in a circle graph.

Instructional Focus: **Challenge**

Data Analysis and Probability: Circle Graphs

	Rebuild Focus	Core 1 Focus	Core 2 Focus	Challenge Focus
	Necklace Chains	**Bag of Chain Links**	**Favorite Colors**	**Circle Graph Plates**
Goal	Construct a circle graph with 2 equal parts	Construct a circle graph with 4 equal parts	Construct a circle graph with unequal parts	Construct a circle graph with complex fractional parts
Materials	♦ Chain links (red and blue) ♦ Poster board, pencil ♦ Red and blue markers	♦ Chain links (4 colors) ♦ Poster board, pencil ♦ Markers (4 colors)	♦ Paper squares (4 colors), tape ♦ Poster board, pencil ♦ Markers (4 colors)	♦ Paper plates ♦ Scissors ♦ Crayons ♦ Strips of paper
Suggested Activity Directions	Show the students a necklace made of 4 blue and 4 red chain links (in a pattern). Explain that a jewelry maker wants to create a circle graph of the links needed to make the necklace. Ask students to unlink the chain links to sort by color and connect the links (1 blue and 1 red). Then connect the 2 links to form a circle (half blue and half red). Trace the circle on the poster board. Draw a line through the center of the circle to show the parts. Students color the sections with the corresponding colors and give the graph a title. Discuss the data. 	Place 20 chain links in a bag (5 of each color). Students randomly take 1 link at a time from the bag. Sort and connect the links as they are chosen. Connect links into a circle. Trace the circle on poster board. Draw lines to show each section. Color and label the circle graph. Disconnect the links and repeat creating another circle graph. Compare the 2 circle graphs. If necessary, turn the poster board to help students see that the graphs are the same. 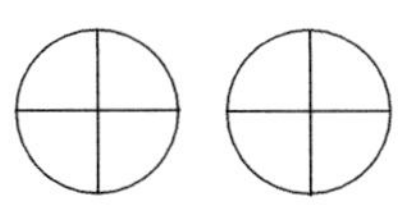	Create a table of data about favorite colors (red = 8, blue = 10, yellow = 2, green = 4). Ask students to construct a circle graph of the table of data. Each piece of data will be represented by a paper square. Prompt discussion about how to sort and organize the data. Model how to put like colors together into a strip and how to tape the strips together to form a cylinder. Trace circle on poster board. Draw lines from the edge of the circle (where colors change on the cylinder) to the center to create a circle graph. Label the graph. Discuss the data. 	Ask students to use paper strips to measure the circumference of a plate and cut the paper strip the same size. Demonstrate how to fold the strip in half, then fifths, creating tenths. Students wrap the paper strip around the edge of the paper plate and draw lines (from each fold) to divide the plate into tenths. Pose this scenario: There are 10 pieces of cake and 4 people. Each person has a different number of slices. Use your strip of paper to figure out how many slices of cake each person has. (Answer: 4, 3, 2, 1) Color and label the sections on the graph. Discuss the percentages.
Questions to Assess	♦ How many of each color are needed? ♦ How much of the circle graph is red?	♦ What do you think the circle graph will look like? ♦ Are the graphs the same? Why?	♦ Why is it important to group the like colors? ♦ Who may like to know this information?	♦ If 4 tenths equal 40%, what percent is 3 tenths? ♦ What percent is the entire circle graph?

Traditional Formative Assessment

More people have three cousins. True or False?

Number of Cousins I Have

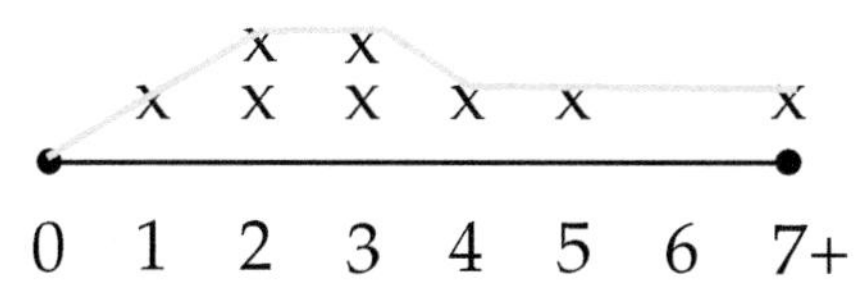

Limitations: In addition to the common limitations of True/False questions, it does not make sense to connect the Xs because this is not a line graph and there is no change over time.

Enhanced Formative Assessment

Add an X on the line plot that represents you.
State at least one fact about the line plot.

Number of Cousins I Have

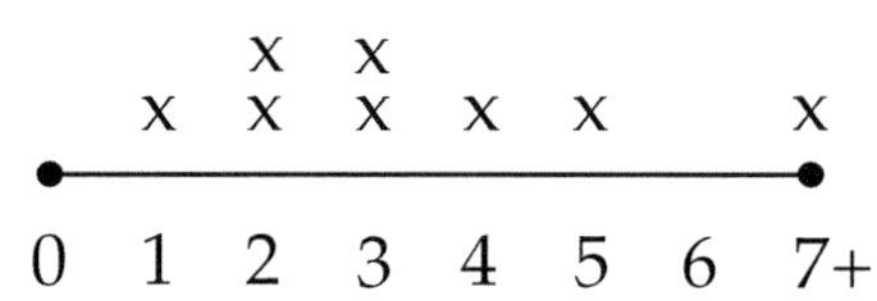

CCSS Data Analysis and Probability: Line Plots

Missy

Add an X on the line plot that represents you.
State at least one fact about the line plot.

Number of Cousins I Have

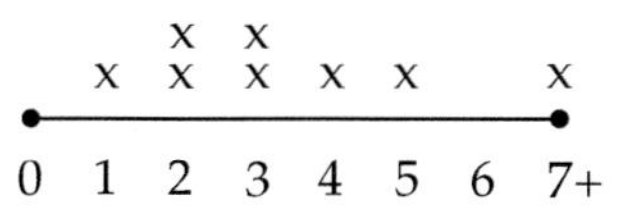

8X

Missy counts the number of Xs on the line plot. She does not accurately place her X on the line plot.

Instructional Focus: **Rebuild**

Nye

Add an X on the line plot that represents you.
State at least one fact about the line plot.

Number of Cousins I Have

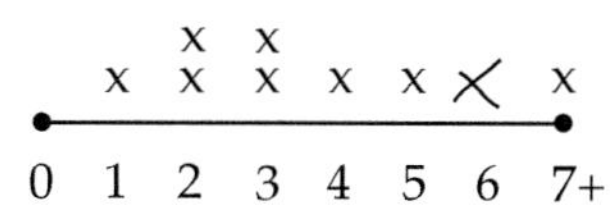

Although Nye places an X on the line plot, he does not put the X in the correct place because he actually has only one cousin.

Instructional Focus: **Core 1**

Oren

Add an X on the line plot that represents you.
State at least one fact about the line plot.

Number of Cousins I Have

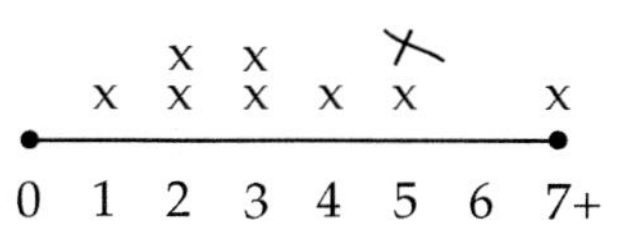

I have five cousins.

Oren accurately places his X on the line plot but does not describe the data.

Instructional Focus: **Core 2**

Pearl

Add an X on the line plot that represents you.
State at least one fact about the line plot.

Number of Cousins I Have

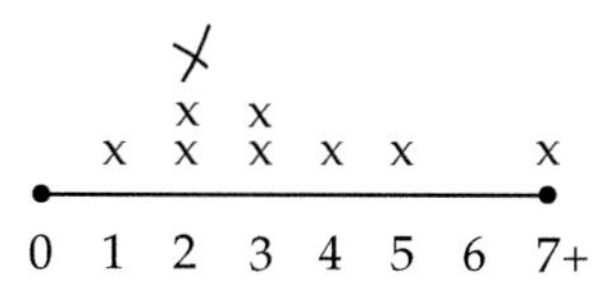

I have two cousins. Most people have two cousins. No one has six or zero cousins.

Pearl correctly places her X on the line plot and offers several facts about the line plot.

Instructional Focus: **Challenge**

CCSS Data Analysis and Probability: Line Plots

	Rebuild Focus	Core 1 Focus	Core 2 Focus	Challenge Focus
	How Many Buttons?	**People in My Family**	**Letters in My Name**	**Handful of Cubes**
Goal	Construct simple line plots with objects	Construct simple line plots with photographs	Describe and analyze line plots	Construct more complex line plots
Materials	♦ Buttons ♦ Poster board ♦ Marker, tape ♦ Paper squares	♦ Photographs of students ♦ Paper, scissors, glue ♦ Poster board, marker	♦ Dry erase boards ♦ Markers	♦ Dry erase boards ♦ Markers ♦ Basket of cubes ♦ Basket of bears
Suggested Activity Directions	Ask students to count the number of buttons on clothing (each person needs at least 1 button). Each student takes the corresponding number of buttons and tapes all of the buttons on 1 paper square. Encourage students to sort the data (buttons on squares) by quantity. Draw a line on the poster board. Label the line 1–10. Help students place their paper squares above the corresponding numbers on the line plot. Discuss the data. NOTE: Each square represents 1 person on the line plot. Buttons taped on the square help bridge from actual objects to representations.	Mount photographs of students on paper rectangles that are the same size and color. Give each student his photograph. Draw a line on the poster board. Ask: "How many people are in your family?" Provide students time to respond. Add numbers to the line plot. Ask students to place each photograph above the corresponding number of people in the family. Discuss the data. Increase the amount of data by asking other class members to participate, if appropriate. 	Draw a line on the dry erase board. Add numbers to the line plot. Ask: "How many letters in your first name?" Students draw an X above the corresponding number. Ask students to draw the line plot on individual dry erase boards. Discuss the data. Try a new line plot with the number of letters in last names. Encourage students to ask one another how many letters and record the data on individual line plots. 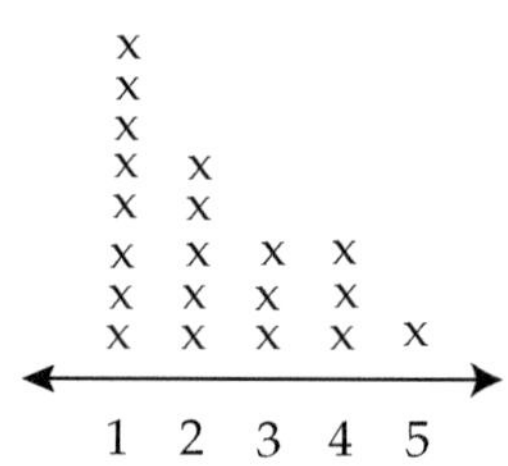	Ask students to predict how many cubes they can grab in one handful. Each student takes 1 handful of cubes. Ask: "How can we make a line plot of these data?" Help students construct the line plot on the dry erase board using an X to represent the number of cubes in each handful. Repeat the activity with bears. Encourage students to construct the new line plot on individual dry erase boards. Compare the line plots. 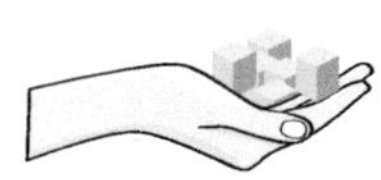
Questions to Assess	♦ How many people have 2 buttons? ♦ Which number of buttons is the most common?	♦ How many people have 4 family members? ♦ Does anyone have 7 family members?	♦ How many letters are in your first name? ♦ Which number of letters is the least common?	♦ Which handful quantity is most common? ♦ Who may want to know this information?

Traditional Formative Assessment

There are 4 green marbles in the bag.

There is 1 blue marble in the bag.

There are 10 red marbles in the bag.

If you take 1 marble, without looking, which color marble will you take?

- ○ Green
- ○ Blue
- ○ Red
- ○ Yellow

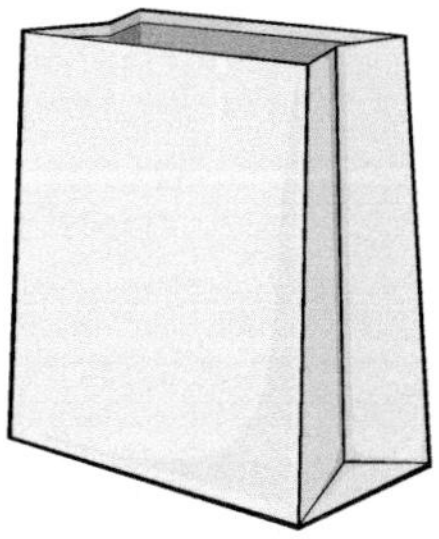

Limitations: In addition to the common problems with multiple-choice questions, the question is worded incorrectly. The question should be, "Which color are you more likely to take?"

Enhanced Formative Assessment

There are 4 green marbles in the bag.
There is 1 blue marble in the bag.
There are 10 red marbles in the bag.

If you take 1 marble without looking, what is the probability of taking red?

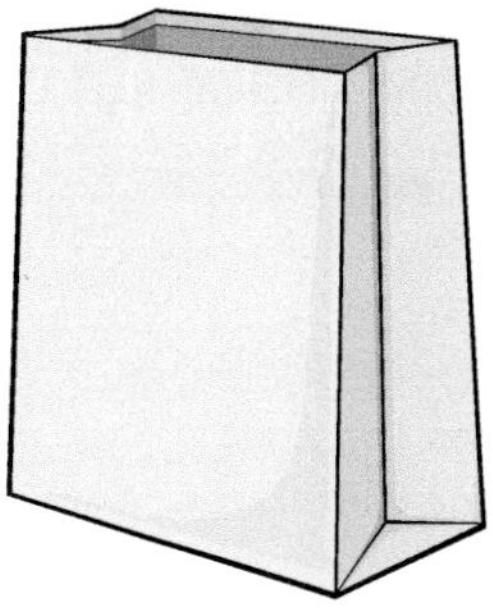

Data Analysis and Probability: Likely or Unlikely Events/Outcomes

Quillan

There are 4 green marbles in the bag.
There is 1 blue marble in the bag.
There are 10 red marbles in the bag.

If you take 1 marble without looking, what is the probability of taking red?

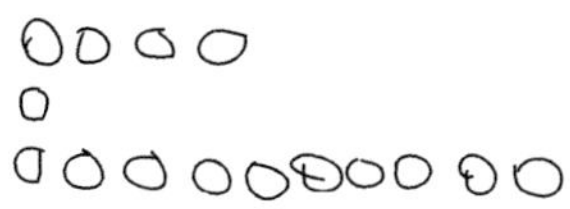

Quillan draws circles to represent the marbles in the bag. He does not record the probability.

Instructional Focus: **Rebuild**

Rocky

There are 4 green marbles in the bag.
There is 1 blue marble in the bag.
There are 10 red marbles in the bag.

If you take 1 marble without looking, what is the probability of taking red?

good

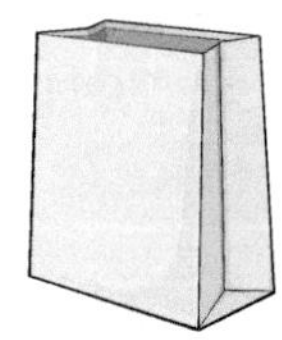

Rocky responds that the probability is "good." He has some understanding but does not use probability vocabulary.

Instructional Focus: **Core 1**

Stacey

There are 4 green marbles in the bag.
There is 1 blue marble in the bag.
There are 10 red marbles in the bag.

If you take 1 marble without looking, what is the probability of taking red?

red is more likely

Stacey's answer is correct. She uses probability vocabulary. She is ready to learn how to describe outcomes with fractions.

Instructional Focus: **Core 2**

Tomi

There are 4 green marbles in the bag.
There is 1 blue marble in the bag.
There are 10 red marbles in the bag.

If you take 1 marble without looking, what is the probability of taking red?

high

10 out of 15

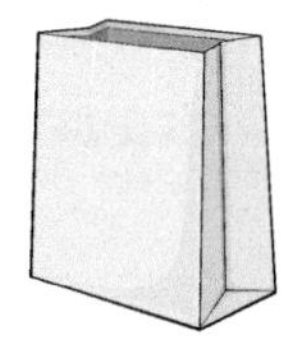

Tomi's answer and explanation are correct. She has a high level of understanding.

Instructional Focus: **Challenge**

Data Analysis and Probability: Likely or Unlikely Events/Outcomes

	Rebuild Focus	Core 1 Focus	Core 2 Focus	Challenge Focus
	Pick a Cube	**Probability Scales**	**Color Spinner**	**Number Spinner**
Goal	Experiment with events that are more likely or less likely	Describe the likelihood of events	Describe probability outcomes with fractions	Describe probability outcomes with percentages
Materials	♦ Cubes (red, blue, green, yellow) ♦ Cups	♦ Pieces of string (same size) ♦ Tape	♦ Spinner with four equal sections (red, blue, yellow, green) ♦ Paper and pencil	♦ Spinner with 10 equal sections (labeled 1–10)
Suggested Activity Directions	Drop 10 red cubes and 1 blue cube in a cup. Cover the cup and shake it to mix the cubes. Ask students which color is more likely to be chosen. One student closes his eyes and chooses a cube. Discuss the outcome. Place the cube back in the cup and repeat giving each student a turn. Encourage students to predict the outcomes using phrases such as, "red is more likely." Ask: "What is the likelihood of choosing blue?" Repeat activity using phrases such as, "blue is less likely." Give each student a cup and cubes. Students place cubes in cups to show more likely (to choose green) and less likely (to choose yellow).	To represent a probability scale, stretch a string straight and tape ends on the table for each student. Ask students to point to the left end of the string. Ask how to describe this position on the probability scale. Encourage students to use the word and the number (impossible, zero). Repeat with the right end of the string (certain, 1). Ask students to point to the place on the scale that represents equally likely (center). Repeat with less likely and more likely. Encourage students to point to locations on the scale for various events such as the likelihood of rain today or of seeing an elephant fly.	Prompt students to analyze the spinner. Discuss the outcomes as equally likely. Ask students how many possible outcomes there are (4). Explain that the outcome of the spinner landing on red is 1 out of 4. Write this as a fraction (1/4). Ask students to describe the other outcomes using fractions. Encourage students to think about the fractional outcome of the spinner landing on blue or green (1/2); yellow or red (1/2); blue, red, or green (3/4); yellow, red, or blue (3/4); yellow, blue, red, or green (4/4). Encourage students to write the fractions as they discuss the outcomes.	Prompt students to analyze the spinner. Discuss the outcomes as equally likely. Ask students how many possible outcomes there are (10). Explain that 1 out of 10 equals 10%. Ask students to figure out what the probability is for the spinner landing on each number (10%). Ask students to figure out the probability of the spinner landing on an even number (50%) and discuss how they determined the answer. Ask students to figure out the probability of the spinner landing on 6 or less (60%), 7 or more (40%), multiple of 3 (30%), and a number less than 11 (100%).
Questions to Assess	♦ Why is red more likely? ♦ How can you make green more likely?	♦ Where is "highly likely" on the probability scale? ♦ How do you know?	♦ What is the likelihood of the spinner landing on green or red?	♦ What is the probability of the spinner landing on an odd number? ♦ How do you know?

7

student
progress

Conclusion

performance
monitoring

"If I only had more time!" This is likely the single greatest request among teachers reflecting on their instruction and the progress of their students. We constantly lament over the lack of time to teach, the lack of time for students to develop conceptual knowledge, and the lack of time to plan effectively for students. So how do we access this valuable commodity? The answer is the implementation of frequent and effective formative assessment. Formative assessment must occur on a regular basis to provide an immediate gauge of student understanding on specific objectives, and the format must be effective in gleaning specific information directing next steps for students.

Assessing and monitoring student performance is a complex task. The purpose of assessment is multifaceted. Formative assessments enable teachers to evaluate the effectiveness of the lesson, provide valuable insight into student learning, and offer a basis for future instruction. The constant administration of formative assessments and the analysis of student responses inform a meaningful and targeted response to the class, to a group of students, or to individuals within the mathematics classroom. The use of formative assessment streamlines the entire instructional cycle. It delineates the difference between teaching a curriculum and teaching students. Teaching solely to the curriculum focuses on a specific standard and assumes all students enter a given mathematics class with the same experiences and background. We know, however, that is not the case. Without formative assessment, we remain unaware of the gaps and misconceptions. Formative assessment allows rediagnosis and reteaching to be done in a timely manner. It is the benefit of meeting students where they are so as to help them progress toward a specific standard. This approach taps into that valuable resource of time because we prevent students from wasting time on things they already know or teaching too far beyond that point which enables them to bridge the gap.

Monitoring student performance through formative assessment enables us to provide a tailored instructional sequence altered to most appropriately facilitate learning. It is our hope that this resource provides examples of effective formative assessments, models of common student responses, and meaningful instructional responses to meet the needs of students. We hope this resource enables you to gain some essential time by identifying and responding to student needs through formative assessment.

Bibliography

Black, P., & Wiliam, D. (1998). Inside the black box: Raising standards through classroom assessment. *Phi Delta Kappa, 80*(2), 139–148.

Common Core State Standards Initiative (2010). http://www.corestandards.org/

Hattie, J., & Jaeger, R. (1998). Assessment and classroom learning: A deductive approach. *Assessment in Education, 5*(1), 111–122.

National Mathematics Advisory Panel. (2008). *Foundations for success: The final report of the National Mathematics Advisory Panel*. Washington, DC: U.S. Department of Education.

National Council of Teachers of Mathematics (NCTM). (2000). *Principles and standards of school mathematics*. Reston, VA: Author.

National Council of Teachers of Mathematics (NCTM). (2006). *Curriculum focal points for prekindergarten through grade 8 mathematics: A quest for coherence*. Reston, VA: Author.

Popham, W. J. (2008). *Transformative assessment*. Alexandria, VA: Association for Supervision and Curriculum Development.

Taylor-Cox, J. (2009). *Math intervention: Building number power with formative assessments, differentiation, and games*. Larchmont, NY: Eye On Education.

Taylor-Cox, J. (2005). *Family math night: Math standards in action*. Larchmont, NY: Eye On Education.

Notes

Notes

Notes

Notes

For Product Safety Concerns and Information please contact our EU representative GPSR@taylorandfrancis.com Taylor & Francis Verlag GmbH, Kaufingerstraße 24, 80331 München, Germany

Batch number: 10733557

Printed by Printforce, the Netherlands